MYWORKBOOK

CHRISTINE VERITY

BASIC COLLEGE MATHEMATICS
NINTH EDITION

Margaret L. Lial
American River College

Stanley A. Salzman
American River College

Diana L. Hestwood
Minneapolis Community and Technical College

PEARSON

Boston   Columbus   Indianapolis   New York   San Francisco   Upper Saddle River
Amsterdam   Cape Town   Dubai   London   Madrid   Milan   Munich   Paris   Montreal   Toronto
Delhi   Mexico City   Sao Paulo   Sydney   Hong Kong   Seoul   Singapore   Taipei   Tokyo

Copyright © 2014, 2010, 2006 Pearson Education, Inc.
Publishing as Pearson, 75 Arlington Street, Boston, MA 02116.

ISBN-13: 978-0-321-83682-3
ISBN-10: 0-321-83682-0

1 2 3 4 5 6 EBM 16 15 14 13 12

www.pearsonhighered.com

CONTENTS

Name: Date:
Instructor: Section:

Chapter 2 MULTIPLYING AND DIVIDING FRACTIONS

2.1   Basics of Fractions

Learning Objectives
1   Use a fraction to show how many parts of a whole are shaded.
2   Identify the numerator and denominator.
3   Identify proper and improper fractions.

Key Terms

Use the vocabulary terms listed below to complete each statement in exercises 1–4.

numerator   **denominator**   **proper fraction**

improper fraction

1.   A fraction whose numerator is greater than or equal to its denominator is called an

_____.

2.   The _____ shows how many equivalent parts are
being considered.

3.   A fraction whose denominator is larger than its numerator is called a

_____.

4.   The _____ of a fraction shows the number of equal
parts in a whole.

Guided Examples

Review these examples for Objective 1:

1.   Use fractions to represent the shaded portions and
the unshaded portions of each figure.

    a.
    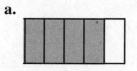

    The figure above has 5 equal parts. The
    1 unshaded part is represented by the

    fraction $\frac{1}{5}$. The shaded part is $\frac{4}{5}$.

Now Try:

1.   Use fractions to represent the
shaded portions and the
unshaded portions of each
figure.

    a.
    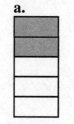

    _____

b.

The figure above has 8 equal parts. The 5 shaded parts are represented by the fraction $\frac{5}{8}$. The unshaded part is $\frac{3}{8}$.

2. Use a fraction to represent the shaded portions the figure.

The area equal to 8 of the $\frac{1}{5}$ parts is shaded, so $\frac{8}{5}$ is shaded.

b.

_____

2. Use a fraction to represent the shaded portions the figure.

_____

Review these examples for Objective 2:

3. Identify the numerator and denominator in each fraction.

    a. $\frac{4}{7}$

    $\frac{4}{7}$ ← Numerator
    ← Denominator

    b. $\frac{9}{4}$

    $\frac{9}{4}$ ← Numerator
    ← Denominator

Now Try:

3. Identify the numerator and denominator in each fraction.

    a. $\frac{5}{6}$

    _____

    b. $\frac{11}{5}$

    _____

Name: Date:
Instructor: Section:

Review these examples for Objective 3:

4. Use this list to classify fractions.

$$\frac{2}{3} \quad \frac{7}{10} \quad \frac{15}{4} \quad \frac{5}{5} \quad \frac{8}{7} \quad \frac{13}{27} \quad \frac{1}{6} \quad \frac{5}{2}$$

    a. Identify all proper fractions in the list.

Proper fractions have a numerator that is less than the denominator. The proper fractions are

$$\frac{2}{3} \quad \frac{7}{10} \quad \frac{13}{27} \quad \frac{1}{6}$$

    b. Identify all improper fractions in the list.

Improper fractions have a numerator that is equal to or greater than the denominator. The improper fractions are

$$\frac{15}{4} \quad \frac{5}{5} \quad \frac{8}{7} \quad \frac{5}{2}$$

Now Try:

4. Use this list to classify fractions.

$$\frac{4}{5} \quad \frac{6}{13} \quad \frac{21}{10} \quad \frac{10}{3} \quad \frac{4}{4} \quad \frac{17}{19} \quad \frac{1}{18} \quad \frac{9}{8}$$

    a. Identify all proper fractions in the list.

_____

    b. Identify all improper fractions in the list.

_____

Objective 1   Use a fraction to show how many parts of a whole are shaded.

For extra help, see Examples 1–2 on page 114 of your text and Section Lecture video for Section 2.1 and Exercise Solutions Clip 5, 9, 21, and 29.

Write the fractions that represent the shaded and unshaded portions of each figure.

1.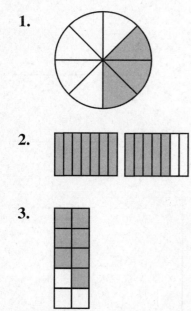

1. Shaded _____

    Unshaded _____

2.

2. Shaded _____

    Unshaded _____

3.

3. Shaded _____

    Unshaded _____

Name: _____   Date: _____

Instructor: _____   Section: _____

Objective 2   Identify the numerator and denominator.

For extra help, see Example 3 on page 115 of your text and Section Lecture video for Section 2.1 and Exercise Solutions Clip 5, 9, 21, and 29.

Identify the numerator and denominator.

4. $\dfrac{8}{11}$

5. $\dfrac{19}{50}$

6. $\dfrac{157}{12}$

4.
Numerator _____

Denominator _____

5.
Numerator _____

Denominator _____

6.
Numerator _____

Denominator _____

Objective 3   Identify proper and improper fractions.

For extra help, see Example 4 on page 115 of your text and Section Lecture video for Section 2.1 and Exercise Solutions Clip 5, 9, 21, and 29.

Write whether each fraction is **proper** *or* **improper**.

7. $\dfrac{17}{11}$

8. $\dfrac{18}{18}$

9. $\dfrac{7}{12}$

7. _____

8. _____

9. _____

Chapter 2 MULTIPLYING AND DIVIDING FRACTIONS

2.2   Mixed Numbers

Learning Objectives
1   Identify mixed numbers.
2   Write mixed numbers as improper fractions.
3   Write improper fractions as mixed numbers.

Key Terms

Use the vocabulary terms listed below to complete each statement in exercises 1–4.

mixed number   **improper fraction**   **proper fraction**

whole numbers

1.   The fraction $\frac{5}{8}$ is an example of a _____ .

2.   A _____ includes a fraction and a whole number written together.

3.   A mixed number can be rewritten as an _____ .

4.   0, 1, 2, 3, … are _____ .

Guided Examples

Review this example for Objective 2:

1.   Write $5\frac{3}{4}$ as an improper fraction.

   Step 1 Multiply the denominator of the fraction and the whole number.

   $5\frac{3}{4}$ $5 \cdot 4 = 20$

   Step 2 Add to this product the numerator of the fraction.

   $5\frac{3}{4}$ $20 + 3 = 23$

   Step 3 Write the result of Step 2 as the numerator of the improper fraction and the original denominator as the denominator.

   $5\frac{3}{4} = \frac{23}{4}$

Now Try:

1.   Write $8\frac{4}{5}$ as an improper fraction.

   _____

Name: Date:
Instructor: Section:

Review these examples for Objective 3: | **Now Try:**
2. Write each improper fraction as a mixed number. | 2. Write each improper fraction as a mixed number.

a. $\dfrac{19}{4}$ | a. $\dfrac{21}{5}$

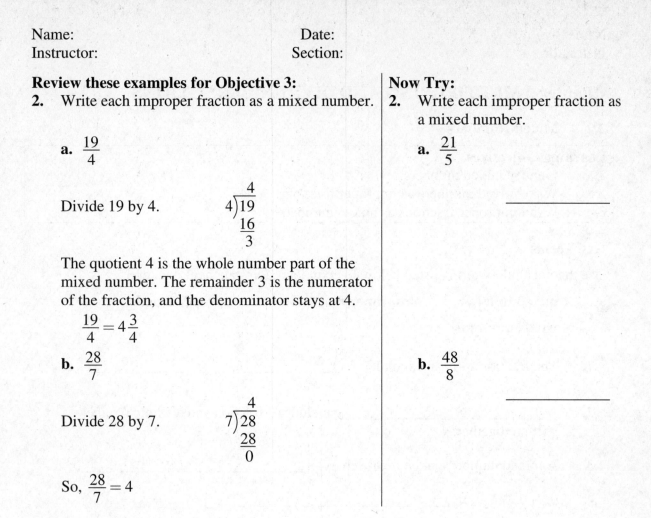

Divide 19 by 4.

$$4\overline{)19}$$

The quotient 4 is the whole number part of the mixed number. The remainder 3 is the numerator of the fraction, and the denominator stays at 4.

$$\frac{19}{4} = 4\frac{3}{4}$$

b. $\dfrac{28}{7}$ | b. $\dfrac{48}{8}$

Divide 28 by 7.

$$7\overline{)28}$$

So, $\dfrac{28}{7} = 4$

Objective 1   Identify mixed numbers.

For extra help, see page 121 of your text and Section Lecture video for Section 2.2.

List the mixed numbers in each group.

1. $2\dfrac{1}{2}, \dfrac{3}{5}, 1\dfrac{1}{6}, \dfrac{3}{4}$ 1. _____

2. $\dfrac{9}{9}, 3\dfrac{1}{2}, 10\dfrac{1}{3}, \dfrac{8}{2}, \dfrac{7}{9}$ 2. _____

3. $\dfrac{6}{3}, 4\dfrac{3}{4}, \dfrac{10}{10}, \dfrac{1}{3}, \dfrac{0}{8}$ 3. _____

Name: Date:
Instructor: Section:

Objective 2   Write mixed numbers as improper fractions.

For extra help, see Example 1 on page 122 of your text and Section Lecture video for Section 2.2 and Exercise Solutions Clip 10 and 25.

Write each mixed number as an improper fraction.

4. $5\frac{4}{7}$ 4. _____

5. $6\frac{1}{4}$ 5. _____

6. $2\frac{7}{11}$ 6. _____

Objective 3   Write improper fractions as mixed numbers.

For extra help, see Example 2 on page 122 of your text and Section Lecture video for Section 2.2 and Exercise Solutions Clip 37, 51, and 59.

Write each improper fraction as a mixed number.

7. $\frac{41}{9}$ 7. _____

8. $\frac{92}{3}$ 8. _____

9. $\frac{211}{11}$ 9. _____

Chapter 2 MULTIPLYING AND DIVIDING FRACTIONS

2.3   Factors

Learning Objectives
1   Find factors of a number.
2   Identify prime numbers and composite numbers.
3   Find prime factorizations.

Key Terms

Use the vocabulary terms listed below to complete each statement in exercises 1–5.

factors   **composite number**   **prime number**

factorizations   **prime factorization**

1.   The numbers that can be multiplied to give a specific number (product) are
     _____ of that number.

2.   A _____ has at least one factor other than itself and 1.

3.   In a _____ every factor is a prime number.

4.   Two different factors of a _____ are itself and 1.

5.   Numbers that are multiplied to give a product are _____.

Guided Examples

Review these examples for Objective 1:

1.   Find all possible two-number factorizations of each numbers.

     a. 24

     $1 \cdot 24 = 24 \quad 2 \cdot 12 = 24 \quad 3 \cdot 8 = 24 \quad 4 \cdot 6 = 24$
     The factors of 24 are 1, 2, 3, 4, 6, 8, 12, and 24.

     b. 48

     $1 \cdot 48 = 44 \qquad 2 \cdot 24 = 48$
     $3 \cdot 16 = 48 \qquad 4 \cdot 12 = 48$
     $6 \cdot 8 = 48$
     The factors of 48 are 1, 2, 3, 4, 6, 8, 12, 16, 24, and 48.

Now Try:

1.   Find all possible two-number factorizations of each numbers.

     a. 32

     _____

     b. 56

     _____

Name: _____ Date: _____

Instructor: _____ Section: _____

Review these examples for Objective 2:

2. Which of the following numbers are prime?
   3, 7, 13, 21, 39

   The number 21 can be divided by 3 and 7, so it is not prime. Also, because 39 can be divided by 3 and 13, then 39 is not prime. The other numbers in the list, 3, 7, and 13 are divisible only by themselves and 1, so they are prime.

3. Which of the following numbers are composite?

   a. 9

   Because 9 has factors of 3, as well as 9 and 1, 9 is composite.

   b. 43

   The number 43 has only two factors, 43 and 1. It is not composite. It is a prime number.

   c. 35

   Because 35 has factors of 5 and 7, as well as 1 and 35, 35 is composite.

Now Try:

2. Which of the following numbers are prime?
   6, 11, 23, 37, 51

   _____

3. Which of the following numbers are composite?

   a. 38

   _____

   b. 31

   _____

   c. 57

   _____

Review these examples for Objective 3:

4. Find the prime factorization of 18.

   Try to divide 18 by the first prime, 2.
   $$18 \div 2 = 9.$$
   so
   $$18 = 2 \cdot 9.$$
   Try to divide 9 by the prime, 3.
   $$9 \div 3 = 3.$$
   so
   $$18 = 2 \cdot 3 \cdot 3.$$
   Because all factors are prime, the prime factorization of 18 is $2 \cdot 3 \cdot 3$.

5. Find the prime factorization of 72.

   $3)\overline{3}^{\,1}$ Continue to divide until the quotient is 1. Divide 3 by 3.
   $3)\overline{9}$ Divide 9 by 3.
   $2)\overline{18}$ Divide 18 by 2.
   $2)\overline{36}$ Divide 36 by 2.
   $2)\overline{72}$ Divide 72 by 2.

   Because all factors (divisors) are prime, the prime factorization of 72 is
   $$72 = 2 \cdot 2 \cdot 2 \cdot 3 \cdot 3 \text{ or } 2^3 \cdot 3^2.$$

Now Try:

4. Find the prime factorization of 24.

   _____

5. Find the prime factorization of 56.

   _____

6. Find the prime factorization of 375.

$\frac{1}{5\overline{)5}}$ Continue to divide until the quotient is 1.
        Divide 5 by 5.

$5\overline{)25}$ Divide 25 by 5.

$5\overline{)125}$ Divide 125 by 5.

$3\overline{)375}$ Divide 375 by 3.

Because all factors (divisors) are prime, the prime factorization of 375 is

$375 = 3 \cdot 5 \cdot 5 \cdot 5$ or $3 \cdot 5^3$.

7. Find the prime factorization of each number using a factor tree.

a. 75

Try to divide by the prime, 3. Write the factors under the 75. Draw a circle (or box) around the 3, since it is prime.

Since 25 cannot be evenly divided by 3, try the next prime, 5.

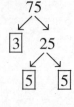

← Circle (or box) because they are prime.

No uncircled (unboxed) factors remain, so the prime factorization has been found.

$75 = 3 \cdot 5 \cdot 5 = 3 \cdot 5^2$

b. 54

Divide by 2.

```
       54
      ↙  ↘
    [2]   27
         ↙  ↘
       [3]   9
            ↙  ↘
          [3]  [3]
```

$54 = 2 \cdot 3 \cdot 3 \cdot 3$, or using exponents, $54 = 2 \cdot 3^3$

6. Find the prime factorization of 150.

_____

7. Find the prime factorization of each number using a factor tree.

a. 36

_____

b. 117

_____

c. 63   c. 40

Because 63 cannot be divided by 2, try 3.   _____

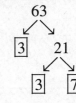

$63 = 3 \cdot 3 \cdot 7$, or using exponents, $63 = 3^2 \cdot 7$

Objective 1   Find factors of a number.

For extra help, see Example 1 on page 126 of your text and Section Lecture video for Section 2.3 and Exercise Solutions Clip 1 and 5.

Find all the factors of each number.

1.   14   1. _____

2.   20   2. _____

3.   72   3. _____

Objective 2   Identify prime numbers and composite numbers.

For extra help, see Examples 2–3 on pages 126–127 of your text and Section Lecture video for Section 2.3 and Exercise Solutions Clip 17, 19, 20 , 25, and 28.

Write whether each number is **prime**, **composite**, *or* **neither**.

4.   24   4. _____

5.   45   5. _____

6.   29   6. _____

Name: Date:
Instructor: Section:

Objective 3 Find prime factorizations.

For extra help, see Examples 4–7 on pages 127–129 of your text and Section Lecture video for Section 2.3 and Exercise Solutions Clip 37 and 48.

Find the prime factorization of each number. Write the answer with exponents when repeated factors appear.

7. 28   **7.** _____

8. 72   **8.** _____

9. 450   **9.** _____

Chapter 2 MULTIPLYING AND DIVIDING FRACTIONS

2.4   Writing a Fraction in Lowest Terms

Learning Objectives
1   Tell whether a fraction is written in lowest terms.
2   Write a fraction in lowest terms using common factors.
3   Write a fraction in lowest terms using prime factors.
4   Determine whether two fractions are equivalent.

Key Terms

Use the vocabulary terms listed below to complete each statement in exercises 1–3.

equivalent fractions   common factor   lowest terms

1.   A fraction is written in _____ when its numerator and denominator have no common factor other than 1.

2.   A _____ is a number that can be divided into two or more whole numbers.

3.   Two fractions are _____ when they represent the same portion of a whole.

Guided Examples

Review these examples for Objective 1:

1.   Are the following fractions in lowest terms?

a. $\dfrac{5}{7}$

The numerator and denominator have no common factor other than 1, so the fraction is in lowest terms.

b. $\dfrac{28}{35}$

The numerator and denominator have a common factor of 7, so the fraction is not in lowest terms.

Now Try:

1.   Are the following fractions in lowest terms?

a. $\dfrac{4}{9}$

_____

b. $\dfrac{6}{15}$

_____

Review these examples for Objective 2:

2. Write each fraction in lowest terms.

    a. $\dfrac{12}{18}$

The greatest common factor of 12 and 18 is 6. Divide both numerator and denominator by 6.

$$\frac{12}{18} = \frac{12 \div 6}{18 \div 6} = \frac{2}{3}$$

    b. $\dfrac{45}{50}$

Divide both numerator and denominator by 5.

$$\frac{45}{50} = \frac{45 \div 5}{50 \div 5} = \frac{9}{10}$$

    c. $\dfrac{32}{48}$

Divide both numerator and denominator by 16.

$$\frac{32}{48} = \frac{32 \div 16}{48 \div 16} = \frac{2}{3}$$

    d. $\dfrac{28}{56}$

Suppose we thought that 4 was the greatest common factor of 28 and 56. Dividing by 4 would give

$$\frac{28}{56} = \frac{28 \div 4}{56 \div 4} = \frac{7}{14}$$

But $\dfrac{7}{14}$ is not in lowest terms, because 7 and 14 have a common factor of 7. So we divide by 7.

$$\frac{7}{14} = \frac{7 \div 7}{14 \div 7} = \frac{1}{2}$$

The fraction $\dfrac{28}{56}$ could have been written in lowest terms in one step by dividing by 28, the greatest common factor of 28 and 56.

$$\frac{28}{56} = \frac{28 \div 28}{56 \div 28} = \frac{1}{2}$$

Now Try:

2. Write each fraction in lowest terms.

    a. $\dfrac{7}{21}$

    _____

    b. $\dfrac{10}{25}$

    _____

    c. $\dfrac{18}{27}$

    _____

    d. $\dfrac{27}{45}$

    _____

Name: Date:
Instructor: Section:

Review these examples for Objective 3:

3. Write each fraction in lowest terms.

    a. $\dfrac{54}{72}$

    Write the prime factorization of both numerator and denominator.

    $$\frac{54}{72} = \frac{3 \cdot 3 \cdot 3 \cdot 2}{3 \cdot 3 \cdot 2 \cdot 2 \cdot 2}$$

    Now divide both numerator and denominator by any common factors. Write a 1 by each factor that has been divided.

    $$\frac{54}{72} = \frac{\overset{1}{\cancel{3}} \cdot \overset{1}{\cancel{3}} \cdot 3 \cdot \overset{1}{\cancel{2}}}{\underset{1}{\cancel{3}} \cdot \underset{1}{\cancel{3}} \cdot 2 \cdot 2 \cdot \underset{1}{\cancel{2}}}$$

    Multiply the remaining factors in both numerator and denominator.

    $$\frac{54}{72} = \frac{1 \cdot 1 \cdot 3 \cdot 1}{1 \cdot 1 \cdot 2 \cdot 2 \cdot 1} = \frac{3}{4}$$

    Finally, $\dfrac{54}{72}$ written in lowest terms is $\dfrac{3}{4}$.

    b. $\dfrac{144}{36}$

    Write the prime factorization of both numerator and denominator.

    $$\frac{144}{36} = \frac{2 \cdot 2 \cdot 2 \cdot 2 \cdot 3 \cdot 3}{2 \cdot 2 \cdot 3 \cdot 3}$$

    Now divide by the common factors. Don't forget to write in the 1s.

    $$\frac{144}{36} = \frac{\overset{1}{\cancel{2}} \cdot \overset{1}{\cancel{2}} \cdot 2 \cdot 2 \cdot \overset{1}{\cancel{3}} \cdot \overset{1}{\cancel{3}}}{\underset{1}{\cancel{2}} \cdot \underset{1}{\cancel{2}} \cdot \underset{1}{\cancel{3}} \cdot \underset{1}{\cancel{3}}}$$

    $$= \frac{1 \cdot 1 \cdot 2 \cdot 2 \cdot 1}{1 \cdot 1 \cdot 1 \cdot 1} = \frac{4}{1} = 4$$

    c. $\dfrac{15}{90}$

    $$\frac{15}{90} = \frac{\overset{1}{\cancel{3}} \cdot \overset{1}{\cancel{5}}}{2 \cdot 3 \cdot \underset{1}{\cancel{3}} \cdot \underset{1}{\cancel{5}}} = \frac{1 \cdot 1}{2 \cdot 3 \cdot 1 \cdot 1} = \frac{1}{6}$$

Now Try:

3. Write each fraction in lowest terms.

    a. $\dfrac{63}{84}$

    _____

    b. $\dfrac{132}{96}$

    _____

    c. $\dfrac{30}{210}$

    _____

Name: Date:

Instructor: Section:

Review these examples for Objective 4:

4. Determine whether each pair of fractions is equivalent. In other words, do both fractions represent the same part of a whole?

a. $\dfrac{30}{45}$ and $\dfrac{32}{48}$

Use the method of prime factors to write each fraction in lowest terms.

$$\frac{30}{45} = \frac{2 \cdot \cancel{3} \cdot \cancel{5}}{3 \cdot \cancel{3} \cdot \cancel{5}} = \frac{2 \cdot 1 \cdot 1}{3 \cdot 1 \cdot 1} = \frac{2}{3}$$

$$\frac{32}{48} = \frac{\cancel{2} \cdot \cancel{2} \cdot \cancel{2} \cdot \cancel{2} \cdot 2}{\cancel{2} \cdot \cancel{2} \cdot \cancel{2} \cdot \cancel{2} \cdot 3} = \frac{1 \cdot 1 \cdot 1 \cdot 1 \cdot 2}{1 \cdot 1 \cdot 1 \cdot 1 \cdot 3} = \frac{2}{3}$$

The fractions are equivalent.

b. $\dfrac{36}{54}$ and $\dfrac{63}{108}$

$$\frac{36}{54} = \frac{\cancel{2} \cdot 2 \cdot \cancel{3} \cdot \cancel{3}}{\cancel{2} \cdot 3 \cdot \cancel{3} \cdot \cancel{3}} = \frac{1 \cdot 2 \cdot 1 \cdot 1}{1 \cdot 3 \cdot 1 \cdot 1} = \frac{2}{3}$$

$$\frac{63}{108} = \frac{\cancel{3} \cdot \cancel{3} \cdot 7}{2 \cdot 2 \cdot 3 \cdot \cancel{3} \cdot \cancel{3}} = \frac{1 \cdot 1 \cdot 7}{2 \cdot 2 \cdot 3 \cdot 1 \cdot 1} = \frac{7}{12}$$

The fractions are not equivalent.

c. $\dfrac{96}{24}$ and $\dfrac{60}{15}$

$$\frac{96}{24} = \frac{\cancel{2} \cdot \cancel{2} \cdot \cancel{2} \cdot 2 \cdot 2 \cdot \cancel{3}}{\cancel{2} \cdot \cancel{2} \cdot \cancel{2} \cdot \cancel{3}} = \frac{1 \cdot 1 \cdot 1 \cdot 2 \cdot 2 \cdot 1}{1 \cdot 1 \cdot 1 \cdot 1} = \frac{4}{1} = 4$$

$$\frac{60}{15} = \frac{2 \cdot 2 \cdot \cancel{3} \cdot \cancel{5}}{\cancel{3} \cdot \cancel{5}} = \frac{2 \cdot 2 \cdot 1 \cdot 1}{1 \cdot 1} = \frac{4}{1}$$

The fractions are equivalent.

Now Try:

4. Determine whether each pair of fractions is equivalent.

a. $\dfrac{35}{5}$ and $\dfrac{140}{20}$

_____

b. $\dfrac{17}{51}$ and $\dfrac{13}{52}$

_____

c. $\dfrac{55}{80}$ and $\dfrac{54}{81}$

_____

Objective 1   Tell whether a fraction is written in lowest terms.

For extra help, see Example 1 on page 132 of your text and Section Lecture video for Section 2.4.

Write whether or not each fraction is in lowest terms. Write **yes** *or* **no**.

1. $\dfrac{12}{18}$ 1. _____

2. $\dfrac{13}{17}$ 2. _____

3. $\dfrac{3}{39}$ 3. _____

Objective 2   Write a fraction in lowest terms using common factors.

For extra help, see Example 2 on page 133 of your text and Section Lecture video for Section 2.4 and Exercise Solutions Clip 17 and 25.

Write each fraction in lowest terms.

4. $\dfrac{14}{49}$ 4. _____

5. $\dfrac{8}{36}$ 5. _____

6. $\dfrac{30}{42}$ 6. _____

Objective 3   Write a fraction in lowest terms using prime factors.

For extra help, see Example 3 on pages 134–135 of your text and Section Lecture video for Section 2.4 and Exercise Solutions Clip 35 and 37.

Write the numerator and denominator of each fraction as a product of prime factors and divide by the common factors. Then write the fraction in lowest terms.

7. $\dfrac{72}{90}$ 7. _____

8. $\dfrac{71}{142}$ 8. _____

9. $\dfrac{75}{500}$ 9. _____

Objective 4   Determine whether two fractions are equivalent.

For extra help, see Example 4 on page 135 of your text and Section Lecture video for Section 2.4 and Exercise Solutions Clip 43.

Determine whether each pair of fractions is **equivalent** *or* **not equivalent**.

10. $\dfrac{8}{16}$ and $\dfrac{15}{20}$ 10. _____

11. $\dfrac{12}{28}$ and $\dfrac{18}{42}$ 11. _____

12. $\dfrac{20}{24}$ and $\dfrac{15}{31}$ 12. _____

Chapter 2 MULTIPLYING AND DIVIDING FRACTIONS

2.5   Multiplying Fractions

Learning Objectives
1   Multiply fractions.
2   Use a multiplication shortcut.
3   Multiply a fraction and a whole number.
4   Find the area of a rectangle.

Key Terms

Use the vocabulary terms listed below to complete each statement in exercises 1–4.

multiplication shortcut   numerator   denominator

common factor

1. A _____ can be divided into two or more whole numbers.

2. The number below the fraction bar in a fraction is called the _____.

3. When multiplying fractions, the process of dividing a numerator and denominator by a common factor can be used as a _____.

4. The number above the fraction bar in a fraction is called the _____.

Guided Examples

Review these examples for Objective 1:

1. Multiply. Write answers in lowest terms.

   a. $\dfrac{6}{7} \cdot \dfrac{4}{5}$

   Multiply the numerators and multiply the denominators.

   $$\dfrac{6}{7} \cdot \dfrac{4}{5} = \dfrac{6 \cdot 4}{7 \cdot 5} = \dfrac{24}{35}$$

   Notice that 24 and 35 have no common factors other than 1, so the answer is in lowest terms.

   b. $\dfrac{9}{10} \cdot \dfrac{3}{7}$

   $$\dfrac{9}{10} \cdot \dfrac{3}{7} = \dfrac{9 \cdot 3}{10 \cdot 7} = \dfrac{27}{70}$$

Now Try:

1. Multiply. Write answers in lowest terms.

   a. $\dfrac{1}{3} \cdot \dfrac{4}{5}$

   _____

   b. $\dfrac{5}{7} \cdot \dfrac{1}{3} \cdot \dfrac{1}{9}$

   _____

Name: Date:
Instructor: Section:

c. $\dfrac{1}{3}\cdot\dfrac{5}{6}\cdot\dfrac{7}{8}$

$\dfrac{1}{3}\cdot\dfrac{5}{6}\cdot\dfrac{7}{8}=\dfrac{1\cdot5\cdot7}{3\cdot6\cdot8}=\dfrac{35}{144}$

c. $\dfrac{1}{5}\cdot\dfrac{3}{10}\cdot\dfrac{7}{20}$

_____

Review these examples for Objective 2:

2. Multiply $\dfrac{4}{5}$ and $\dfrac{15}{16}$. Write answers in lowest terms.

$\dfrac{4}{5}\cdot\dfrac{15}{16}=\dfrac{4\cdot15}{5\cdot16}=\dfrac{60}{80}$ ← Not in lowest terms.

The numerator and denominator have a common factor other than 1, so write the prime factorization of each number.

$\dfrac{4}{5}\cdot\dfrac{15}{16}=\dfrac{4\cdot15}{5\cdot16}=\dfrac{2\cdot2\cdot3\cdot5}{5\cdot2\cdot2\cdot2\cdot2}$

Next, divide by the common factors of 2, 2, and 5.

$\dfrac{4}{5}\cdot\dfrac{15}{16}=\dfrac{4\cdot15}{5\cdot16}=\dfrac{\overset{1}{\cancel{2}}\cdot\overset{1}{\cancel{2}}\cdot3\cdot\overset{1}{\cancel{5}}}{\underset{1}{\cancel{5}}\cdot\underset{1}{\cancel{2}}\cdot\underset{1}{\cancel{2}}\cdot2\cdot2}$

Finally, multiply the remaining factors in the numerator and in the denominator.

$\dfrac{4}{5}\cdot\dfrac{15}{16}=\dfrac{1\cdot1\cdot3\cdot1}{1\cdot1\cdot1\cdot2\cdot2}=\dfrac{3}{4}$

As a shortcut, instead of writing the prime factorization of each number, find the product of $\dfrac{4}{5}$ and $\dfrac{15}{16}$ as follows.

First, divide by 4, a common factor of both 4 and 16.

$\dfrac{\overset{1}{\cancel{4}}}{5}\cdot\dfrac{15}{\underset{4}{\cancel{16}}}$

Next, divide by 5, a common factor of both 5 and 15.

$\dfrac{\overset{1}{\cancel{4}}}{\underset{1}{\cancel{5}}}\cdot\dfrac{\overset{3}{\cancel{15}}}{\underset{4}{\cancel{16}}}$

Finally, multiply numerators and multiply denominators.

$\dfrac{1\cdot3}{1\cdot4}=\dfrac{3}{4}$

Now Try:

2. Multiply $\dfrac{9}{17}$ and $\dfrac{34}{27}$. Write answers in lowest terms.

_____

84   Copyright © 2014 Pearson Education, Inc.

3. Use the multiplication shortcut to find each product. Write the answers in lowest terms and as mixed numbers where possible.

a. $\dfrac{5}{17} \cdot \dfrac{34}{35}$

Divide both 5 and 35 by their common factor of 5. Then divide 17 and 34 by their common factor of 17.

$$\dfrac{\overset{1}{\cancel{5}}}{\underset{1}{\cancel{17}}} \cdot \dfrac{\overset{2}{\cancel{34}}}{\underset{7}{\cancel{35}}} = \dfrac{1 \cdot 2}{1 \cdot 7} = \dfrac{2}{7}$$

b. $\dfrac{7}{12} \cdot \dfrac{24}{35}$

Divide 7 and 35 by 7, then divide 12 and 24 by 12.

$$\dfrac{\overset{1}{\cancel{7}}}{\underset{1}{\cancel{12}}} \cdot \dfrac{\overset{2}{\cancel{24}}}{\underset{5}{\cancel{35}}} = \dfrac{1 \cdot 2}{1 \cdot 5} = \dfrac{2}{5}$$

c. $\dfrac{9}{4} \cdot \dfrac{16}{15}$

$$\dfrac{\overset{3}{\cancel{9}}}{\underset{1}{\cancel{4}}} \cdot \dfrac{\overset{4}{\cancel{16}}}{\underset{5}{\cancel{15}}} = \dfrac{3 \cdot 4}{1 \cdot 5} = \dfrac{12}{5} \text{ or } 2\dfrac{2}{5}$$

d. $\dfrac{3}{4} \cdot \dfrac{5}{9} \cdot \dfrac{2}{5}$

$$\dfrac{\overset{1}{\cancel{3}}}{\underset{2}{\cancel{4}}} \cdot \dfrac{\overset{1}{\cancel{5}}}{\underset{3}{\cancel{9}}} \cdot \dfrac{\overset{1}{\cancel{2}}}{\underset{1}{\cancel{5}}} = \dfrac{1 \cdot 1 \cdot 1}{2 \cdot 3 \cdot 1} = \dfrac{1}{6}$$

Review these examples for Objective 3:

4. Multiply. Write answers in lowest terms and as whole numbers where possible.

a. $6 \cdot \dfrac{2}{3}$

Write 6 as $\dfrac{6}{1}$ and multiply.

$$6 \cdot \dfrac{2}{3} = \dfrac{\overset{2}{\cancel{6}}}{1} \cdot \dfrac{2}{\underset{1}{\cancel{3}}} = \dfrac{2 \cdot 2}{1 \cdot 1} = \dfrac{4}{1} = 4$$

3. Use the multiplication shortcut to find each product. Write the answers in lowest terms and as mixed numbers where possible.

a. $\dfrac{3}{5} \cdot \dfrac{25}{27}$

_____

b. $\dfrac{5}{6} \cdot \dfrac{4}{35}$

_____

c. $\dfrac{15}{7} \cdot \dfrac{28}{9}$

_____

d. $\dfrac{25}{36} \cdot \dfrac{14}{35} \cdot \dfrac{3}{7}$

_____

Now Try:

4. Multiply. Write answers in lowest terms and as whole numbers where possible.

a. $9 \cdot \dfrac{2}{9}$

_____

b. $18 \cdot \dfrac{7}{8}$

b. $16 \cdot \dfrac{5}{8}$

_____

$$18 \cdot \frac{7}{8} = \frac{\overset{9}{\cancel{18}}}{1} \cdot \frac{7}{\underset{4}{\cancel{8}}} = \frac{9 \cdot 7}{1 \cdot 4} = \frac{63}{4} = 15\frac{3}{4}$$

Review these examples for Objective 4:	**Now Try:**

5. Find the area of each rectangle, multiply its length by its width.

5. Find the area of each rectangle, multiply its length by its width.

a. Length is $\dfrac{3}{4}$ mi; width is $\dfrac{2}{5}$ mi

a. Length is $\dfrac{6}{7}$ yd;

width is $\dfrac{7}{12}$ yd

Area = length · width

Area $= \dfrac{3}{4} \cdot \dfrac{2}{5}$

_____

$$= \frac{3}{\underset{2}{\cancel{4}}} \cdot \frac{\overset{1}{\cancel{2}}}{5}$$

$$= \frac{3}{10} \text{ square mi (mi}^2)$$

b. Length is $\dfrac{9}{25}$ in.; width is $\dfrac{5}{18}$ in.

b. Length is $\dfrac{6}{5}$ m;

Area $= \dfrac{9}{25} \cdot \dfrac{5}{18}$

width is $\dfrac{25}{3}$ m

$$= \frac{\overset{1}{\cancel{9}}}{\underset{5}{\cancel{25}}} \cdot \frac{\overset{1}{\cancel{5}}}{\underset{2}{\cancel{18}}}$$

_____

$$= \frac{1}{10} \text{ square in. (in}^2)$$

Objective 1   Multiply factions.

For extra help, see Example 1 on page 143 of your text and Section Lecture video for Section 2.5.

Multiply. Write answers in lowest terms.

1. $\dfrac{3}{4} \cdot \dfrac{5}{6} \cdot \dfrac{2}{3}$

1. _____

2. $\dfrac{9}{10} \cdot \dfrac{3}{2}$

2. _____

3. $\dfrac{1}{9} \cdot \dfrac{2}{3} \cdot \dfrac{5}{6}$ 3. _____

Objective 2   Use a multiplication shortcut.

For extra help, see Examples 2–3 on pages 143–145 of your text and Section Lecture video for Section 2.5 and Exercise Solutions Clip 9, 14, and 18.

Use the multiplication shortcut to find each product. Write the answer in lowest terms.

4. $\dfrac{7}{6} \cdot \dfrac{3}{14}$ 4. _____

5. $\dfrac{11}{4} \cdot \dfrac{8}{33}$ 5. _____

6. $\dfrac{3}{8} \cdot \dfrac{4}{9} \cdot \dfrac{15}{6}$ 6. _____

Objective 3   Multiply a fraction and a whole number.

For extra help, see Example 4 on page 145–146 of your text and Section Lecture video for Section 2.5 and Exercise Solutions Clip 22 and 24.

Multiply. Write the answer in lowest terms. Change the answer to a whole or mixed number where possible.

7. $49 \cdot \dfrac{6}{7}$ 7. _____

8. $27 \cdot \dfrac{7}{54}$ 8. _____

9. $200 \cdot \dfrac{7}{50} \cdot \dfrac{5}{28}$ 9. _____

Objective 4   Find the area of a rectangle.

For extra help, see Example 5 on pages 146–147 of your text and Section Lecture video for Section 2.5 and Exercise Solutions Clip 37.

Find the area of each rectangle.

10.   Length: $\frac{2}{3}$ yard, width: $\frac{1}{2}$ yard

10. _____

11.   Length: $\frac{5}{3}$ meters, width: $\frac{3}{2}$ meters

11. _____

Solve the application problem.

12.   A desk is $\frac{2}{3}$ yard by $\frac{5}{6}$ yard. Find its area.

12. _____

Name: Date:
Instructor: Section:

Chapter 2 MULTIPLYING AND DIVIDING FRACTIONS

2.6   Applications of Multiplication

Learning Objectives
1   Solve fraction application problems using multiplication.

Key Terms

Use the vocabulary terms listed below to complete each statement in exercises 1–3.

   reciprocals   product   indicator words

1.   The words "times" and "double" are _____ for multiplication.

2.   In the problem $51 \times 3 = 153$, 153 is called the _____.

3.   Two numbers are _____ of each other if their product is 1.

Guided Examples

Review these examples for Objective 1:
1.   Lani paid $120 for textbooks this term. Of this amount, the bookstore kept $\frac{1}{4}$. How much did the bookstore keep?

Step 1 Read the problem. The problem asks us to find the amount of money the bookstore kept.

Step 2 Work out a plan. The indicator word is of. The word of indicates multiplication, so find the amount the bookstore kept by multiplying $\frac{1}{4}$ and $120.

Step 3 Estimate a reasonable answer. Round $120 to $100 and divide by 4. Our estimate is $100 \div 4 = \$25$.

Step 4 Solve the problem.

$$\text{amount} = \frac{1}{\overset{}{\underset{1}{4}}} \cdot \overset{30}{\cancel{120}} = \frac{30}{1} = 30$$

Step 5 State the answer. The bookstore kept $30.

Step 6 Check. The exact answer, $30, is close to our estimate of $25.

Now Try:
1.   A store sells 3750 items, of which $\frac{2}{15}$ are classified as junk food. How many of the items are junk food?

   _____

2. Of the 570 employees of Grand Tire Service, $\frac{7}{30}$ have given to the United Fund. How many have given to the United Fund?

Step 1 Read the problem. The problem asks us to find the number of employees who have given to the United Fund.

Step 2 Work out a plan. The indicator word is of. The word of indicates multiplication, so find the number of employees by multiplying $\frac{7}{30}$ and 570.

Step 3 Estimate a reasonable answer. Round 570 to 600. Then, $\frac{1}{4}$ of 600 is 150. Since $\frac{7}{30}$ is less than $\frac{1}{4}$, our estimate is that "less than 150 employees" have given to the United Fund.

Step 4 Solve the problem.

$$\text{amount} = \frac{7}{\cancel{30}_{1}} \cdot \cancel{570}^{19} = \frac{7 \cdot 19}{1} = 133$$

Step 5 State the answer. 133 employees have given to the United Fund.

Step 6 Check. The exact answer, 133, fits our estimate of "less than 150 employees."

3. During the month of April, $\frac{5}{9}$ of the days had temperatures that were below normal. Of the days that had below normal temperatures, $\frac{3}{4}$ were rainy. What fraction of the days in April were rainy and below normal in temperature?

Step 1 Read the problem. The problem asks for the fraction of days that were rainy and below normal in temperature.

Step 2 Work out a plan. Reword the problem to read "in April $\frac{5}{9}$ of $\frac{3}{4}$ days were rainy and below normal temperatures."

Step 3 Estimate a reasonable answer. Since $\frac{5}{9}$ is

2. Deepak puts $\frac{1}{12}$ of his weekly earnings in a retirement fund. If he makes $1248 a week, how much does he put in his retirement fund each week?

_____

3. Last week when Dan delivered pizzas, he noticed that $\frac{1}{4}$ of his customers gave him exact change. Of the customers who had exact change, $\frac{2}{5}$ gave him a tip. What fraction of the customers with exact change gave him a tip?

_____

more than $\frac{1}{2}$, and $\frac{1}{2} \cdot \frac{3}{4} = \frac{3}{8}$, our estimate is

"more than $\frac{3}{8}$."

Step 4 Solve the problem.

$$\text{amount} = \frac{5}{\overset{6}{\underset{3}{\cancel{6}}}} \cdot \frac{\overset{1}{\cancel{3}}}{4} = \frac{5 \cdot 1}{3 \cdot 4} = \frac{5}{12}$$

Step 5 State the answer. $\frac{5}{12}$ of the days in April were rainy and below normal temperatures.

Step 6 Check. The exact answer, $\frac{5}{12}$, fits our estimate of "more than $\frac{3}{8}$."

Objective 1   Solve fraction application problems using multiplication.

For extra help, see Examples 1–4 on pages 152–154 of your text and Section Lecture video for Section 2.6 and Exercise Solutions Clip 11, 15a, and 25.

Solve each application problem.

1. The Donut Shack sells donuts, bagels, and muffins. During a typical week, they sell 1120 items, of which $\frac{2}{7}$ are muffins. How many muffins does the Donut Shack sell in a typical week?

    1. _____

2. Major league baseball teams play 162 games during the regular season. If a team wins $\frac{15}{27}$ of its games, how many games does it win?

    2. _____

3. Elena noticed that her gas gauge moved from the $\frac{7}{8}$ mark to the $\frac{4}{8}$ mark during a recent trip. If her tank holds 16 gallons, how many gallons did she use during this trip?

    3. _____

Name: Date:
Instructor: Section:

Chapter 2 MULTIPLYING AND DIVIDING FRACTIONS

2.7   Dividing Fractions

Learning Objectives
1   Find the reciprocal of a fraction.
2   Divide fractions.
3   Solve application problems in which fractions are divided.

Key Terms

Use the vocabulary terms listed below to complete each statement in exercises 1–3.

reciprocals   indicator words   quotient

1. Two numbers are _____ of each other if their product is 1.

2. The words "per" and "divided equally" are _____ for division.

3. In the problem $192 \div 12 = 16$, 16 is called the _____.

Guided Examples

Review these examples for Objective 1:
1. Find the reciprocal of each fraction.

   a. $\frac{1}{9}$

   The reciprocal of $\frac{1}{9}$ is $\frac{9}{1}$ because $\frac{1}{9} \cdot \frac{9}{1} = \frac{9}{9} = 1$

   b. $\frac{3}{7}$

   The reciprocal of $\frac{3}{7}$ is $\frac{7}{3}$ because $\frac{3}{7} \cdot \frac{7}{3} = \frac{21}{21} = 1$

   c. $\frac{5}{11}$

   The reciprocal of $\frac{5}{11}$ is $\frac{11}{5}$ because

   $\frac{5}{11} \cdot \frac{11}{5} = \frac{55}{55} = 1$

   d. 6

   The reciprocal of 6 is $\frac{1}{6}$ because $\frac{6}{1} \cdot \frac{1}{6} = \frac{6}{6} = 1$

Now Try:
1. Find the reciprocal of each fraction.

   a. $\frac{1}{6}$

   _____

   b. $\frac{4}{13}$

   _____

   c. $\frac{12}{25}$

   _____

   d. 21

   _____

Copyright © 2014 Pearson Education, Inc.

Name: _____   Date: _____

Instructor: _____   Section: _____

Review these examples for Objective 2:

2. Divide. Write answers in lowest terms and as mixed numbers where possible.

   a. $\dfrac{5}{6} \div \dfrac{25}{18}$

   The reciprocal of $\dfrac{25}{18}$ is $\dfrac{18}{25}$.

   Change division to multiplication and use the reciprocal.

   $$\dfrac{5}{6} \div \dfrac{25}{18} = \dfrac{5}{6} \cdot \dfrac{18}{25}$$

   $$= \dfrac{\overset{1}{\cancel{5}}}{\underset{1}{\cancel{6}}} \cdot \dfrac{\overset{3}{\cancel{18}}}{\underset{5}{\cancel{25}}} = \dfrac{1 \cdot 3}{1 \cdot 5} = \dfrac{3}{5}$$

   b. $\dfrac{\frac{4}{7}}{\frac{5}{14}}$

   $$\dfrac{\frac{4}{7}}{\frac{5}{14}} = \dfrac{4}{7} \div \dfrac{5}{14} = \dfrac{4}{\underset{1}{\cancel{7}}} \cdot \dfrac{\overset{2}{\cancel{14}}}{5} = \dfrac{4 \cdot 2}{1 \cdot 5} = \dfrac{8}{5} = 1\dfrac{3}{5}$$

3. Divide. Write answers in lowest terms and as whole or mixed numbers where possible.

   a. $4 \div \dfrac{1}{7}$

   Write 4 as $\dfrac{4}{1}$. Next, use the reciprocal of $\dfrac{1}{7}$, which is $\dfrac{7}{1}$.

   $$4 \div \dfrac{1}{7} = \dfrac{4}{1} \cdot \dfrac{7}{1} = \dfrac{4 \cdot 7}{1 \cdot 1} = \dfrac{28}{1} = 28$$

   b. $\dfrac{5}{8} \div 15$

   Write 15 as $\dfrac{15}{1}$. The reciprocal of $\dfrac{15}{1}$ is $\dfrac{1}{15}$.

   $$\dfrac{5}{8} \div 15 = \dfrac{5}{8} \cdot \dfrac{1}{15} = \dfrac{\overset{1}{\cancel{5}}}{8} \cdot \dfrac{1}{\underset{3}{\cancel{15}}} = \dfrac{1 \cdot 1}{8 \cdot 3} = \dfrac{1}{24}$$

Now Try:

2. Divide. Write answers in lowest terms and as mixed numbers where possible.

   a. $\dfrac{8}{15} \div \dfrac{3}{10}$

   _____

   b. $\dfrac{\frac{4}{9}}{\frac{16}{27}}$

   _____

3. Divide. Write answers in lowest terms and as whole or mixed numbers where possible.

   a. $16 \div \dfrac{1}{5}$

   _____

   b. $\dfrac{11}{3} \div 5$

   _____

Review these examples for Objective 3:

4. Amanda wants to make doll dresses to sell at a craft's fair. Each dress needs $\frac{2}{3}$ yard of material. She has 18 yards of material. Find the number of dresses that she can make.

Step 1 Read the problem. We need to find the number of dresses with $\frac{2}{3}$-yard of material Amanda can make from 18 yards of material.

Step 2 Work out a plan. We solve the problem by finding the number of times $\frac{2}{3}$ goes into 18.

Step 3 Estimate a reasonable answer. Round $\frac{2}{3}$ yard to 1 yard. She would use 1 yard of material 18 times, so our estimate is 18 dresses.

Step 4 Solve the problem.

$$18 \div \frac{2}{3} = \frac{18}{1} \cdot \frac{3}{2} = \frac{\overset{9}{\cancel{18}}}{1} \cdot \frac{3}{\underset{1}{\cancel{2}}} = \frac{27}{1} = 27$$

Step 5 State the answer. Amanda can make 27 dresses with 18 yards of material.

Step 6 Check. The exact answer, 27 dresses, is reasonably close to our estimate of 18 dresses.

5. Abel has a piece of property with an area of $\frac{7}{8}$ acre. He wishes to divide it into four equal parts for his children. How many acres of land will each child get?

Step 1 Read the problem. Since $\frac{7}{8}$ acre of land must be split into four parts, we must find the fraction of land received by each child.

Step 2 Work out a plan. Divide the fraction of land $\left(\frac{7}{8}\right)$ by the number of children (4).

Step 3 Estimate a reasonable answer. Round $\frac{7}{8}$ acre to 1 acre. If the land (1 whole) were divided

Now Try:

4. How many $\frac{1}{9}$-ounce medicine vials can be filled with 7 ounces of medicine?

_____

5. A human resource advisor spends $\frac{2}{3}$ of her day in individual conferences. She must meet with 18 candidates. What fraction of her conference time is spent with each candidate?

_____

between 4 children, each would receive $\frac{1}{4}$ acre, our estimate.

Step 4 Solve the problem.

$$\frac{7}{8} \div 4 = \frac{7}{8} \div \frac{4}{1} = \frac{7}{8} \cdot \frac{1}{4} = \frac{7}{32}$$

Step 5 State the answer. Each child receives $\frac{7}{32}$ acre of the land.

Step 6 Check. The exact answer, $\frac{7}{32}$-acre, is close to our estimate of $\frac{1}{4}$ acre.

Objective 1 Find the reciprocal of a fraction.

For extra help, see Example 1 on page 159 of your text and Section Lecture video for Section 2.7 and Exercise Solutions Clip 5, 9, and 11.

Find the reciprocal of each fraction.

1. $\frac{9}{2}$

1. _____

2. $\frac{1}{3}$

2. _____

3. 10

3. _____

Objective 2 Divide fractions

For extra help, see Examples 2–3 on pages 160–162 of your text and Section Lecture video for Section 2.7 and Exercise Solutions Clip 13, 21, 23, and 25.

Divide. Write the answer in lowest terms. Change the answers to a whole or mixed number where possible.

4. $\frac{28}{5} \div \frac{42}{25}$

4. _____

5. $\frac{\frac{4}{9}}{\frac{16}{27}}$

5. _____

6. $9 \div \dfrac{3}{2}$ 6. _____

Objective 3   Solve application problems in which fractions are divided.

For extra help, see Examples 4–5 on pages 162–163 of your text and Section Lecture video for Section 2.7 and Exercise Solutions Clip 11, 15a, and 25.

Solve each application problem.

7.   Lynn has 2 gallons of lemonade. If each of her   7. _____
         Brownies gets $\frac{1}{12}$ gallon of lemonade, how many
         Brownies does she have?

8.   Bill wishes to make hamburger patties that weight   8 _____
         $\frac{5}{12}$ pound. How many hamburger patties can he
         make with 10 pounds of hamburger?

9.   Glen has a small pickup truck that will carry $\frac{3}{4}$ cord   9. _____
         of firewood. Find the number of trips needed to
         deliver 30 cords of wood.

Chapter 2 MULTIPLYING AND DIVIDING FRACTIONS

2.8   Multiplying and Dividing Mixed Numbers

Learning Objectives
1   Estimate the answer and multiply mixed numbers.
2   Estimate the answer and divide mixed numbers.
3   Solve application problems with mixed numbers.

Key Terms

Use the vocabulary terms listed below to complete each statement in exercises 1–3.

   mixed number   **simplify**   **round**

1.   To _____a fraction means to write the fraction in lowest terms.

2.   If the numerator of a fraction is half of the denominator or more, _____ up to the next whole number to estimate the product of a mixed number and a whole number.

3. $2\frac{7}{11}$ is an example of a _____.

Guided Examples

Review these examples for Objective 1:

1.   First estimate the answer. Then multiply to get an exact answer. Simplify your answers.

   a. $3\frac{1}{2}\cdot 4\frac{2}{7}$

   Estimate the answer by rounding the mixed numbers.

   $3\frac{1}{2}$ rounds to 4 and $4\frac{2}{7}$ rounds to 4

   $4\cdot 4=16$ Estimated answer

   To find the exact answer, change each mixed number to an improper fraction.

   Step 1 Change to improper fractions.

   $3\frac{1}{2}=\frac{7}{2}$ and $4\frac{2}{7}=\frac{30}{7}$

   So, $3\frac{1}{2}\cdot 4\frac{2}{7}=\frac{7}{2}\cdot\frac{30}{7}$

   Step 2 Multiply as fractions.

Now Try:

1.   First estimate the answer. Then multiply to get an exact answer. Simplify your answers.

   a. $5\frac{2}{5}\cdot 6\frac{5}{6}$

   _____

$$= \frac{\overset{1}{\cancel{7}}}{\underset{1}{\cancel{2}}} \cdot \frac{\overset{15}{\cancel{30}}}{\underset{1}{\cancel{7}}} = \frac{1 \cdot 15}{1 \cdot 1}$$

Step 3 Simplify the answer.

$$= \frac{15}{1} = 15$$

The estimated answer is 16 and the exact answer is 15. The exact answer is reasonable.

b. $2\frac{5}{6} \cdot 5\frac{1}{4}$

Estimate the answer by rounding the mixed numbers.

$2\frac{5}{6}$ rounds to 3 and $5\frac{1}{4}$ rounds to 5

$3 \cdot 5 = 15$ Estimated answer

Now find the exact answer.

Step 1

$$2\frac{5}{6} = \frac{17}{6} \text{ and } 5\frac{1}{4} = \frac{21}{4}$$

So, $2\frac{5}{6} \cdot 5\frac{1}{4} = \frac{17}{6} \cdot \frac{21}{4}$

Step 2

$$= \frac{17}{\underset{2}{\cancel{6}}} \cdot \frac{\overset{7}{\cancel{21}}}{4} = \frac{17 \cdot 7}{2 \cdot 4} = \frac{119}{8}$$

Step 3

$$= \frac{119}{8} = 14\frac{7}{8}$$

The estimate was 15, so the exact answer of $14\frac{7}{8}$ is reasonable.

c. $5\frac{3}{4} \cdot 6\frac{2}{3}$

$5\frac{3}{4}$ rounds to 6 and $6\frac{2}{3}$ rounds to 7

$6 \cdot 7 = 42$ Estimated answer

The exact answer is shown below.

$$5\frac{3}{4} \cdot 6\frac{2}{3} = \frac{23}{\underset{1}{\cancel{4}}} \cdot \frac{\overset{5}{\cancel{20}}}{3} = \frac{23 \cdot 5}{1 \cdot 3} = \frac{115}{3} = 38\frac{1}{3}$$

The estimate was 42, so the exact answer of $38\frac{1}{3}$ is reasonable.

b. $2\frac{4}{9} \cdot 3\frac{7}{8}$

_____

c. $1\frac{7}{11} \cdot 4\frac{1}{9}$

_____

Name: Date:

Instructor: Section:

Review these examples for Objective 2:

2. First estimate the answer. Then divide to find the exact answer. Simplify your exact answers.

    a. $3\frac{4}{5} \div 4\frac{1}{3}$

First estimate the answer by rounding each mixed number to the nearest whole number.

$3\frac{4}{5}$ rounds to 4 and $4\frac{1}{3}$ rounds to 4

$4 \div 4 = 1$ Estimated answer

To find the exact answer, first change each mixed number to an improper fraction.

Step 1 Change to improper fractions.

$3\frac{4}{5} = \frac{19}{5}$ and $4\frac{1}{3} = \frac{13}{3}$

So, $3\frac{4}{5} \div 4\frac{1}{3} = \frac{19}{5} \div \frac{13}{3}$

Step 2 Use the reciprocal of the divisor.

$\frac{19}{5} \div \frac{13}{3} = \frac{19}{5} \cdot \frac{3}{13}$

Step 3 Change division to multiplication.

$= \frac{19}{5} \cdot \frac{3}{13} = \frac{19 \cdot 3}{5 \cdot 13}$

Step 3 Simplify the answer.

$= \frac{57}{65}$

The estimate was 1, so the exact answer of $\frac{57}{65}$ is reasonable.

    b. $14 \div 6\frac{3}{5}$

14 rounds to 14 and $6\frac{3}{5}$ rounds to 7

$14 \div 7 = 2$ Estimated answer

The exact answer is shown below.

$14 \div 6\frac{3}{5} = \frac{14}{1} \div \frac{33}{5} = \frac{14}{1} \cdot \frac{5}{33} = \frac{70}{33} = 2\frac{4}{33}$

The estimate was 2, so the exact answer of $2\frac{4}{33}$ is reasonable.

Now Try:

2. First estimate the answer. Then divide to find the exact answer. Simplify your exact answers.

    a. $20\frac{1}{4} \div 2\frac{1}{3}$

        _____

    b. $15 \div 2\frac{7}{8}$

        _____

c. $8\frac{3}{4} \div 5$

$8\frac{3}{4}$ rounds to 9 and 5 rounds to 5

$9 \div 5 = \frac{9}{1} \div \frac{5}{1} = \frac{9}{1} \cdot \frac{1}{5} = \frac{9}{5} = 1\frac{4}{5}$ Estimated

The exact answer is shown below.

$8\frac{3}{4} \div 5 = \frac{35}{4} \div \frac{5}{1} = \frac{35}{4} \cdot \frac{1}{5} = \frac{\overset{7}{\cancel{35}}}{4} \cdot \frac{1}{\underset{1}{\cancel{5}}} = \frac{7}{4} = 1\frac{3}{4}$

The estimate was $1\frac{4}{5}$, so the exact answer of $1\frac{3}{4}$

is reasonable.

c. $6\frac{4}{5} \div 7$

_____

Review these examples for Objective 3:

3. Juan worked $38\frac{1}{4}$ hours at \$9 per hour. How

much did he make?

Step 1 Read the problem. The problem asks for

the wages Juan earned for $38\frac{1}{4}$ hours of work.

Step 2 Work out a plan. Multiply the number of

hours $\left(38\frac{1}{4}\right)$ and the wage per hour (\$9).

Step 3 Estimate a reasonable answer. Round

$38\frac{1}{4}$ hours to 38 hours. Multiply 38 hours by \$9

per hour $(38 \cdot 9)$ to get an estimate of \$342.

Step 4 Solve the problem.

$38\frac{1}{4} \cdot 9 = \frac{153}{4} \cdot \frac{9}{1} = \frac{1377}{4} = 344\frac{1}{4}$

Step 5 State the answer. Juan receives $\$344\frac{1}{4}$

or \$344.25.

Step 6 Check. The exact answer, $\$344\frac{1}{4}$, is

close to our estimate of \$342.

Now Try:

3. Maria wants to make 20 dresses
to sell at a bazaar. Each dress

needs $3\frac{1}{4}$ yards of material.

How many yards does she need?

_____

4. A farmer applies fertilizer to his fields at a rate of $5\frac{5}{6}$ gallons per acre. How many acres can he fertilize with $65\frac{5}{6}$ gallons?

Step 1 Read the problem. The problem asks how many acres can be fertilized with $65\frac{5}{6}$ gallons.

Step 2 Work out a plan. Divide the number of acres $\left(65\frac{5}{6}\right)$ by the rate $\left(5\frac{5}{6}\right)$.

Step 3 Estimate a reasonable answer. Round $65\frac{5}{6}$ to 66 and round $5\frac{5}{6}$ to 6. Divide 66 by 6 $(66 \div 6)$ to get an estimate of 11 acres.

Step 4 Solve the problem.

$$65\frac{5}{6} \div 5\frac{5}{6} = \frac{395}{6} \div \frac{35}{6} = \frac{\overset{79}{\cancel{395}}}{\underset{1}{\cancel{6}}} \cdot \frac{\overset{1}{\cancel{6}}}{\underset{7}{\cancel{35}}} = \frac{79}{7} = 11\frac{2}{7}$$

Step 5 State the answer. $11\frac{2}{7}$ acres can be fertilized with $65\frac{5}{6}$ gallons.

Step 6 Check. The exact answer, $11\frac{2}{7}$ acres, is close to our estimate of 11 acres.

4. How many dresses can be made from 70 yards of material if each dress requires $4\frac{3}{8}$ yards?

_____

Objective 1   Estimate the answer and multiply mixed numbers.

For extra help, see Example 1 on pages 170–171 of your text and Section Lecture video for Section 2.8 and Exercise Solutions Clip 7 and 9.

First estimate the answer. Then multiply to find the exact answer. Simplify all answers.

1. $5\frac{1}{3} \cdot 2\frac{1}{2}$

1.
Estimate _____

Exact _____

2. $4\frac{4}{9} \cdot 2\frac{2}{5}$

2.
Estimate _____

Exact _____

3. $5\frac{2}{3} \cdot 7\frac{1}{8}$

3.
Estimate_____

Exact _____

Objective 2   Estimate the answer and divide mixed numbers.

For extra help, see Example 2 on pages 172–173 of your text and Section Lecture video for Section 2.8 and Exercise Solutions Clip 21.

First estimate the answer. Then divide to find the exact answer. Simplify all answers.

4. $4\frac{5}{8} \div 1\frac{1}{4}$

4.
Estimate_____

Exact _____

5. $4\frac{3}{8} \div 3\frac{1}{2}$

5.
Estimate_____

Exact _____

6. $2\frac{5}{8} \div 1\frac{3}{4}$

6.
Estimate_____

Exact _____

Objective 3   Solve application problems with mixed numbers.

For extra help, see Examples 3–4 on pages 173–174 of your text and Section Lecture video for Section 2.8 and Exercise Solutions Clip 39.

First estimate the answer. Then solve each application problem. Simplify all answers.

7.   Each home in an area needs $41\frac{1}{3}$ yards of rain gutter. How much rain gutter would be needed for 6 homes?

7.
Estimate_____

Exact _____

8.   Arnette worked $24\frac{1}{2}$ hours and earned \$9 per hour. How much did she earn?

8.
Estimate_____

Exact _____

9.   A dental office plays taped music constantly. Each tape takes $1\frac{1}{4}$ hours. How many tapes are played during $7\frac{1}{2}$ hours?

9.
Estimate_____

Exact _____

Chapter 3 ADDING AND SUBTRACTING FRACTIONS

3.1   Adding and Subtracting Like Fractions

Learning Objectives
1   Define like and unlike fractions.
2   Add like fractions.
3   Subtract like fractions.

Key Terms

Use the vocabulary terms listed below to complete each statement in exercises 1–2.

   like fractions   **unlike fractions**

   1.   Fractions with different denominators are called _____.

   2.   Fractions with the same denominator are called _____.

Guided Examples

Review these examples for Objective 1:
1.   Next to each pair of fractions, write like or unlike.

   a. $\frac{9}{7}$, $\frac{2}{7}$

   The fractions $\frac{9}{7}$ and $\frac{2}{7}$ are like fractions since the denominators are the same.

   b. $\frac{3}{5}$, $\frac{4}{10}$

   The fractions $\frac{3}{5}$ and $\frac{4}{10}$ are unlike fractions since the denominators are different.

Now Try:
1.   Next to each pair of fractions, write like or unlike.

   a. $\frac{2}{9}$, $\frac{5}{9}$

   _____

   b. $\frac{2}{3}$, $\frac{3}{2}$

   _____

Review these examples for Objective 2:
2.   Add and write the sum in lowest terms.

   a. $\frac{3}{7}+\frac{2}{7}$

   $\frac{3}{7}+\frac{2}{7}=\frac{3+2}{7}=\frac{5}{7}$ ← Sum of numerators
   ← Same denominator

Now Try:
2.   Add and write the sum in lowest terms.

   a. $\frac{4}{9}+\frac{3}{9}$

   _____

b. $\dfrac{3}{9}+\dfrac{1}{9}+\dfrac{2}{9}$

Step 1 $\quad\dfrac{3+1+2}{9}$

Step 2 $\quad=\dfrac{6}{9}$

Step 3 $\quad=\dfrac{6\div 3}{9\div 3}=\dfrac{2}{3}$

b. $\dfrac{3}{10}+\dfrac{1}{10}+\dfrac{4}{10}$

_____

Review these examples for Objective 3:

3. Find the difference and simplify the answer.

a. $\dfrac{3}{10}-\dfrac{1}{10}$

Step 1 $\quad\dfrac{3}{10}-\dfrac{1}{10}=\dfrac{3-1}{10}$

Step 2 $\qquad\quad=\dfrac{2}{10}\begin{array}{l}\leftarrow\text{Difference of numerators}\\\leftarrow\text{Same denominator}\end{array}$

Step 3 $\qquad\quad=\dfrac{2\div 2}{10\div 2}=\dfrac{1}{5}\leftarrow\text{Lowest terms}$

b. $\dfrac{21}{17}-\dfrac{3}{17}$

$\dfrac{21}{17}-\dfrac{3}{17}=\dfrac{21-3}{17}\begin{array}{l}\leftarrow\text{Difference of numerators}\\\leftarrow\text{Same denominator}\end{array}$

$\qquad\qquad=\dfrac{18}{17}$

To simplify the answer, write $\dfrac{18}{17}$ as a mixed number.

$\dfrac{18}{17}=1\dfrac{1}{17}$

Now Try:

3. Find the difference and simplify the answer.

a. $\dfrac{16}{21}-\dfrac{2}{21}$

_____

b. $\dfrac{24}{15}-\dfrac{6}{15}$

_____

Objective 1   Define like and unlike fractions.

For extra help, see Example 1 on page 198 of your text and Section Lecture video for Section 3.1.

Next to each pair of fractions, write **like** *or* **unlike**.

1. $\quad\dfrac{3}{15},\dfrac{1}{5}$

1. _____

2. $\quad\dfrac{5}{6},\dfrac{1}{6}$

2. _____

3. $\dfrac{6}{11}, \dfrac{9}{11}$

3. _____

Objective 2   Add like fractions.

For extra help, see Example 2 on page 199 of your text and Section Lecture video for Section 3.1 and Exercise Solutions Clip 9, 10, and 17.

Find the sum and simplify the answer.

4. $\dfrac{11}{16} + \dfrac{7}{16}$

4. _____

5. $\dfrac{11}{15} + \dfrac{1}{15}$

5. _____

6. $\dfrac{67}{81} + \dfrac{29}{81} + \dfrac{12}{81}$

6. _____

Objective 3   Subtract like fractions.

For extra help, see Example 3 on page 200 of your text and Section Lecture video for Section 3.1 and Exercise Solutions Clip 19, 27, 31, and 39.

Find the difference and simplify the answer.

7. $\dfrac{25}{28} - \dfrac{15}{28}$

7. _____

8. $\dfrac{31}{36} - \dfrac{11}{36}$

8. _____

9. $\dfrac{91}{20} - \dfrac{41}{20}$

9. _____

Chapter 3 ADDING AND SUBTRACTING FRACTIONS

3.2   Least Common Multiples

Learning Objectives
1   Find the least common multiple (LCM).
2   Find the least common multiple using multiples of the largest number.
3   Find the least common multiple using prime factorization.
4   Find the least common multiple using an alternative method.
5   Write a fraction with an indicated denominator.

Key Terms

Use the vocabulary terms listed below to complete each statement in exercises 1–2.

least common multiple   LCM

1.   Given two or more whole numbers, the _____ is
      the smallest whole number divisible by both of the numbers.

2.   _____ is the abbreviation for least common multiple.

Guided Examples

Review this example for Objective 1:
1.   Find the least common multiple of 4 and 6.

   First, find the multiples of 4.

   $\underbrace{4 \cdot 1}_{4}$, $\underbrace{4 \cdot 2}_{8}$, $\underbrace{4 \cdot 3}_{12}$, $\underbrace{4 \cdot 4}_{16}$, $\underbrace{4 \cdot 5}_{20}$, $\underbrace{4 \cdot 6}_{24}$, $\underbrace{4 \cdot 7}_{28}$, $\underbrace{4 \cdot 8}_{32}$,...

   Now, find the multiples of 6.

   $\underbrace{6 \cdot 1}_{6}$, $\underbrace{6 \cdot 2}_{12}$, $\underbrace{6 \cdot 3}_{18}$, $\underbrace{6 \cdot 4}_{24}$, $\underbrace{6 \cdot 5}_{30}$, $\underbrace{6 \cdot 6}_{36}$, $\underbrace{6 \cdot 7}_{42}$, $\underbrace{6 \cdot 8}_{48}$,...

   The smallest number found in both lists is 12, so
   12 is the least common multiple of 4 and 6; the
   number 12 is the smallest whole number
   divisible by both 4 and 6.

   Multiples of 4: 4, 8, 12, 16, 20, 24, 28, 32,...
                         ↙↗
   Multiples of 6: 6, 12, 18, 24, 30, 36, 42, 48,...

   12 is the smallest number found in both lists. 12
   is the least common multiple (LCM) of 4 and 6.

Now Try:
1.   Find the least common multiple
      of 8 and 14.

      _____

Review this example for Objective 2:
2.   Use multiples of the larger number to find the
      least common multiple of 8 and 12.

   Start by writing the first few multiples of 12.
      Multiples of 12: 12, 24, 36, 48, 60, 72, …

Now Try:
2.   Use multiples of the larger
      number to find the least
      common multiple of 9 and 12.

      _____

Now, check each multiple of 12 to see if it is
divisible by 8.
    12, 24, 36, 48, 60, 72, …
First multiple divisible by 8 is 24, because
$24 \div 8 = 3$.
The least common multiple of 8 and 12 is 24.

Review these examples for Objective 3:	**Now Try:**
3. Use prime factorization to find the least common multiple of 6 and 15.	3. Use prime factorization to find the least common multiple of 12 and 15.

Start by finding the prime factorization of each
number.
$$6 = 2 \cdot 3$$
$$15 = 3 \cdot 5$$
Circle (or box) the factors where they appear the
greatest number of times in either factorization.
$$6 = \boxed{2} \cdot \boxed{3}$$
$$15 = 3 \cdot \boxed{5}$$
The LCM is the product of the circled factors.

$$\text{Factors of 6}$$
$$\downarrow\downarrow$$
$$\text{LCM} = 2 \cdot 3 \cdot 5 = 30$$
$$\uparrow\uparrow$$
$$\text{Factors of 15}$$

The product of the prime factors, 30 is the least
common multiple. The smallest whole number
divisible by both 6 and 15 is 30.

4. Find the least common multiple of 10, 15, and
   18.

4. Find the least common multiple
   of 7, 35, and 15.

Find the prime factorization of each number.
Then use the prime factors to build the LCM.
$$10 = \boxed{2} \cdot 5$$
$$15 = 3 \cdot \boxed{5}$$
$$18 = 2 \cdot \boxed{3} \cdot \boxed{3}$$
$$\text{LCM} = 2 \cdot 3 \cdot 3 \cdot 5 = 90$$
Check to see that 90 is divisible by 10 (yes) and
by 15 (yes) and by 18 (yes). The smallest whole
number divisible by 10, 15, and 18 is 90. The
LCM is 90.

Name: Date:
Instructor: Section:

5. Find the least common multiple for each set of
    numbers.

    a. 6, 7, 28

    Find the prime factorization for each number.
    $6 = 2 \cdot \boxed{3}$
    $7 = \boxed{7}$
    $28 = \boxed{2} \cdot \boxed{2} \cdot 7$
    LCM $= 2 \cdot 2 \cdot 3 \cdot 7 = 84$
    The least common multiple of 6, 7, and 28 is 84.

    b. 12, 16, 18

    Find the prime factorization for each number.
    $12 = 2 \cdot 2 \cdot 3$
    $16 = 2 \cdot 2 \cdot 2 \cdot 2$
    $18 = 2 \cdot 3 \cdot 3$
    LCM $= 2 \cdot 2 \cdot 2 \cdot 2 \cdot 3 \cdot 3 = 144$
    The least common multiple of 12, 16, and 18 is
    144.

5. Find the least common multiple
    for each set of numbers.

    a. 5, 10, 15

    _____

    b. 3, 6, 15

    _____

Review these examples for Objective 4:

6. Find the least common multiple for each set of
    numbers.

    a. 21 and 28

    Start by trying to divide 21 and 28 by the first
    prime number, which is 2. Use the following
    shortcut.

    2 | 2̸1̸   28
    ────────────
       21   14

    Because 21 cannot be divided evenly by 2, cross
    out 21 and bring it down. Divide by 2 again.

    2 | 2̸1̸   28
    2 | 2̸1̸   14
    ────────────
       21   7

    Divide by 3.

    2 | 2̸1̸   28
    2 | 2̸1̸   14
    3 | 21   7̸
    ────────────
       7   7

    Since 7 cannot be divided evenly by the third
    prime, 5, skip 5 and divide by the next prime, 7.

Now Try:

6. Find the least common multiple
    for each set of numbers.

    a. 18 and 30

    _____

Copyright © 2014 Pearson Education, Inc.

```
2 | 2̶1̶   28
2 | 2̶1̶   14
3 | 21    7̶
7 |  7    7
      1    1
```

When all quotients are 1, multiply the prime
numbers on the left side.

Least common multiple $= 2 \cdot 2 \cdot 3 \cdot 7 = 84$

The least common multiple of 21 and 28 is 84.

b. 9, 12, 15   **b.** 24, 36, 9

Divide by 2.

```
2 | 9̶    12    1̶5̶
    9     6    15
```

Divide by 2 again.

```
2 | 9̶    12    1̶5̶
2 | 9̶     6    1̶5̶
    9     3    15
```

Divide by 3.

```
2 | 9̶    12    1̶5̶
2 | 9̶     6    1̶5̶
3 | 9     3    15
    3     1     5
```

Divide by 3 again.

```
2 | 9̶    12    1̶5̶
2 | 9̶     6    1̶5̶
3 | 9     3    15
3 | 3     1̶     5̶
    1     1     5
```

Finally, divide by 5.

```
2 | 9̶    12    1̶5̶
2 | 9̶     6    1̶5̶
3 | 9     3    15
3 | 3     1̶     5̶
5 | 1̶     1̶     5
    1     1     1
```

Multiply the prime numbers on the left side.

Least common multiple $= 2 \cdot 2 \cdot 3 \cdot 3 \cdot 5 = 180$

The least common multiple of 9, 12, and 15 is
180.

7. Find the least common multiple of 15, 18, and 24. Use the (a) prime factorization method and then use (b) the alternative method.

a. Use the prime factorization method.

Find the prime factorization for each number.

$$15 = 3 \cdot \boxed{5}$$
$$18 = 2 \cdot \boxed{3} \cdot \boxed{3}$$
$$24 = \boxed{2} \cdot \boxed{2} \cdot \boxed{2} \cdot 3$$

The product of the boxed factors is the LCM.

$$LCM = 2 \cdot 2 \cdot 2 \cdot 3 \cdot 3 \cdot 5 = 360$$

b. Use the alternative method.

$$
\begin{array}{r|ccc}
2 & \cancel{15} & 18 & 24 \\
2 & \cancel{15} & 9 & 12 \\
2 & \cancel{15} & \cancel{9} & 6 \\
3 & 15 & \cancel{9} & 3 \\
3 & \cancel{5} & 3 & \cancel{1} \\
5 & 5 & \cancel{1} & \cancel{1} \\
& 1 & 1 & 1
\end{array}
$$

$$LCM = 2 \cdot 2 \cdot 2 \cdot 3 \cdot 3 \cdot 5 = 360$$

7. Find the least common multiple of 4, 6, and 15.

a. Use the prime factorization method.

_____

b. Use the alternative method.

_____

Review these examples for Objective 5:

8. Write the fraction $\frac{5}{6}$ with a denominator of 30.

Find a numerator, so that these fractions are equivalent.

$$\frac{5}{6} = \frac{?}{30}$$

To find the new numerator, first divide 30 by 6.

$$\frac{5}{6} = \frac{?}{30} \qquad 30 \div 6 = 5$$

Multiply both numerator and denominator of the fraction $\frac{5}{6}$ by 5.

$$\frac{5}{6} = \frac{5 \cdot 5}{6 \cdot 5} = \frac{25}{30}$$

This process is just the opposite of writing a fraction in lowest terms. Check the answer by writing $\frac{25}{30}$ in lowest terms; you should get $\frac{5}{6}$ again.

Now Try:

8. Write the fraction $\frac{7}{8}$ with a denominator of 40.

_____

Name: Date:

Instructor: Section:

9. Rewrite each fraction with the indicated denominator.

a. $\dfrac{3}{7} = \dfrac{?}{42}$

Divide 42 by 7, getting 6. Now multiply both the numerator and the denominator of $\dfrac{3}{7}$ by 6.

$$\dfrac{3}{7} = \dfrac{3 \cdot 6}{7 \cdot 6} = \dfrac{18}{42}$$

That is, $\dfrac{3}{7} = \dfrac{18}{42}$. As a check write $\dfrac{18}{42}$ in lowest terms; you should get $\dfrac{3}{7}$ again.

b. $\dfrac{11}{12} = \dfrac{?}{60}$

Divide 60 by 12, getting 5. Now multiply both the numerator and the denominator of $\dfrac{11}{12}$ by 5.

$$\dfrac{11}{12} = \dfrac{11 \cdot 5}{12 \cdot 5} = \dfrac{55}{60}$$

This shows that $\dfrac{11}{12} = \dfrac{55}{60}$. As a check write $\dfrac{55}{60}$ in lowest terms; you should get $\dfrac{11}{12}$ again.

9. Rewrite each fraction with the indicated denominator.

a. $\dfrac{7}{15} = \dfrac{?}{45}$

_____

b. $\dfrac{9}{13} = \dfrac{?}{39}$

_____

Objective 1   Find the least common multiple (LCM).

For extra help, see Example 1 on page 205 of your text and Section Lecture video for Section 3.2.

Find the least common multiple for each of the following by listing multiples of each number.

1.   7, 14   **1.** _____

2.   12, 18   **2.** _____

3.   30, 75   **3.** _____

Objective 2   Find the least common multiple using multiples of the largest number.

For extra help, see Example 2 on page 206 of your text and Section Lecture video for Section 3.2 and Exercise Solutions Clip 13.

Find the least common multiple for each of the following by using multiples of the largest number.

    4.   5, 12   **4.** _____

    5.   14, 35   **5.** _____

    6.   32, 40   **6.** _____

Objective 3   Find the least common multiple using prime factorization.

For extra help, see Examples 3–5 on pages 206–207 of your text and Section Lecture video for Section 3.2 and Exercise Solutions Clip 16, 18, and 22.

Find the least common multiple for each of the following using prime factorization.

    7.   14, 48   **7.** _____

    8.   10, 24, 32   **8.** _____

    9.   16, 20, 25   **9.** _____

Objective 4   Find the least common multiple using an alternative method.

For extra help, see Examples 6–7 on page 208–209 of your text and Section Lecture video for Section 3.2 and Exercise Solutions Clip 16, 18, and 22.

Find the least common multiple for each of the following using an alternative method.

10.   22, 55   10. _____

11.   4, 18, 27   11. _____

12.   12, 30, 40   12. _____

Objective 5   Write a fraction with an indicated denominator.

For extra help, see Examples 8–9 on pages 209–210 of your text and Section Lecture video for Section 3.2 and Exercise Solutions Clip 39 and 43.

Rewrite each fraction with the indicated denominator.

13. $\dfrac{1}{9} = \dfrac{}{36}$ 13. _____

14. $\dfrac{1}{13} = \dfrac{}{78}$ 14. _____

15. $\dfrac{15}{7} = \dfrac{}{84}$ 15. _____

Chapter 3 ADDING AND SUBTRACTING FRACTIONS

3.3   Adding and Subtracting Unlike Fractions

Learning Objectives
1   Add unlike fractions.
2   Add unlike fractions vertically.
3   Subtract unlike fractions.
4   Subtract unlike fractions vertically.

Key Terms

Use the vocabulary terms listed below to complete each statement in exercises 1–2.

   least common denominator   LCD

   1.   In order to add or subtract fractions with different denominators, first find the

         _____.

   2.   _____ is the abbreviation for least common denominator.

Guided Examples

Review these examples for Objective 1:

1.   Add $\frac{1}{6}$ and $\frac{5}{18}$.

The least common multiple of 6 and 18 is 18, so first rewrite the fractions as like fractions with a denominator of 18. This is the least common denominator (LCD) of 6 and 18.

Step 1 $\frac{1}{6} = \frac{?}{18}$

Divide 18 by 6, getting 3. Next, multiply numerator and denominator by 3.

$$\frac{1}{6} = \frac{1 \cdot 3}{6 \cdot 3} = \frac{3}{18}$$

Now, add the like fractions $\frac{3}{18}$ and $\frac{5}{18}$.

Step 2 $\frac{1}{6} + \frac{5}{18} = \frac{3}{18} + \frac{5}{18} = \frac{3+5}{18} = \frac{8}{18}$

Step 3 $\frac{8}{18} = \frac{4}{9}$

Now Try:

1.   Add $\frac{5}{8}$ and $\frac{3}{16}$.

         _____

2. Add each pair of fractions using the three steps. Simplify all answers.

    a. $\dfrac{1}{7} + \dfrac{1}{14}$

    The least common multiple of 7 and 14 is 14. Rewrite both fractions as fractions with a least common denominator of 14.

    Step 1 $\dfrac{1}{7} + \dfrac{1}{14} = \dfrac{2}{14} + \dfrac{1}{14}$

    Step 2 $\dfrac{2}{14} + \dfrac{1}{14} = \dfrac{2+1}{14} = \dfrac{3}{14}$

    Step 3 Step 3 is not needed because $\dfrac{3}{14}$ is already written in lowest terms.

    b. $\dfrac{6}{15} + \dfrac{7}{20}$

    The least common multiple of 15 and 20 is 60. Rewrite both fractions as fractions with a least common denominator of 60.

    Step 1 $\dfrac{6}{15} + \dfrac{7}{20} = \dfrac{24}{60} + \dfrac{21}{60}$

    Step 2 $\dfrac{24}{60} + \dfrac{21}{60} = \dfrac{24+21}{60} = \dfrac{45}{60}$

    Step 3 $\dfrac{45}{60} = \dfrac{3}{4}$

2. Add each pair of fractions using the three steps. Simplify all answers.

    a. $\dfrac{1}{4} + \dfrac{1}{12}$

    _____

    b. $\dfrac{5}{9} + \dfrac{1}{6}$

    _____

Review these examples for Objective 2:

3. Add the following fractions vertically.

    a. $\dfrac{3}{14}$

        $+\dfrac{5}{21}$

    Rewrite as like fractions. Then add the numerators.

    $$\dfrac{3}{14} = \dfrac{3\cdot 3}{14\cdot 3} = \dfrac{9}{42}$$

    $$+\dfrac{5}{21} = \dfrac{5\cdot 2}{21\cdot 2} = +\dfrac{10}{42}$$

    $$\dfrac{19}{42}$$

    The denominator is 42, the LCD.

Now Try:

3. Add the following fractions vertically.

    a. $\dfrac{5}{8}$

        $+\dfrac{7}{24}$

    _____

b. $\dfrac{3}{8}$

$+\dfrac{5}{18}$

Rewrite as like fractions. Then add the numerators.

$$\dfrac{3}{8} = \dfrac{3 \cdot 9}{8 \cdot 9} = \dfrac{27}{72}$$

$$+\dfrac{5}{18} = \dfrac{5 \cdot 4}{18 \cdot 4} = +\dfrac{20}{72}$$

$$\dfrac{47}{72}$$

The denominator is 72, the LCD.

b. $\dfrac{5}{12}$

$+\dfrac{2}{9}$

_____

Review these examples for Objective 3:

4. Subtract. Simplify all answers.

  a. $\dfrac{5}{9} - \dfrac{1}{3}$

As with addition, rewrite unlike fractions with a least common denominator.

Step 1 $\dfrac{5}{9} - \dfrac{1}{3} = \dfrac{5}{9} - \dfrac{3}{9}$

Step 2 $\dfrac{5}{9} - \dfrac{3}{9} = \dfrac{5-3}{9} = \dfrac{2}{9}$

Step 3   Not needed because $\dfrac{2}{9}$ is in lowest terms.

  b. $\dfrac{5}{6} - \dfrac{1}{2}$

Step 1 $\dfrac{5}{6} - \dfrac{1}{2} = \dfrac{5}{6} - \dfrac{3}{6}$

Step 2 $\dfrac{5}{6} - \dfrac{3}{6} = \dfrac{5-3}{6} = \dfrac{2}{6}$

Step 3 $\dfrac{2}{6} = \dfrac{1}{3}$

Now Try:

4. Subtract. Simplify all answers.

  a. $\dfrac{5}{7} - \dfrac{3}{5}$

_____

  b. $\dfrac{4}{5} - \dfrac{2}{15}$

_____

Review these examples for Objective 4:

5. Subtract the following fractions vertically. Simplify all answers.

  a. $\dfrac{7}{9}$

$-\dfrac{3}{7}$

Rewrite as like fractions. Then subtract the numerators.

Now Try:

5. Subtract the following fractions vertically. Simplify all answers.

  a. $\dfrac{4}{5}$

$-\dfrac{2}{3}$

_____

$$\frac{7}{9} = \frac{7 \cdot 7}{9 \cdot 7} = \frac{49}{63}$$

$$-\frac{3}{7} = \frac{3 \cdot 9}{7 \cdot 9} = -\frac{27}{63}$$

$$\frac{22}{63}$$

The denominator is 63, the LCD.

$\frac{22}{63}$ is in simplest form.

b. $\frac{7}{12}$

$-\frac{4}{7}$

b. $\frac{5}{9}$

$-\frac{5}{18}$

_____

Rewrite as like fractions. Then subtract the numerators.

$$\frac{7}{12} = \frac{7 \cdot 7}{12 \cdot 7} = \frac{49}{84}$$

$$-\frac{4}{7} = \frac{4 \cdot 12}{7 \cdot 12} = -\frac{48}{84}$$

$$\frac{1}{84}$$

The denominator is 84, the LCD.

$\frac{1}{84}$ is in simplest form.

Objective 1   Add unlike fractions.

For extra help, see Examples 1–2 on pages 215-216 of your text and Section Lecture video for Section 3.3 and Exercise Solutions Clip 7, 11, and 16.

Add the following fractions. Simplify all answers.

1. $\frac{1}{5} + \frac{5}{8}$

1. _____

2. $\frac{4}{15} + \frac{9}{20}$

2. _____

3. $\frac{1}{3} + \frac{1}{8} + \frac{5}{12}$

3. _____

Name: Date:

Instructor: Section:

Objective 2   Add unlike fractions vertically.

For extra help, see Example 3 on page 217 of your text and Section Lecture video for Section 3.3 and Exercise Solutions Clip 19 and 21.

Add the following fractions. Simplify all answers.

4. $\dfrac{1}{15}$

$+\dfrac{2}{3}$

4. _____

5. $\dfrac{5}{22}$

$+\dfrac{7}{33}$

5. _____

6. $\dfrac{6}{13}$

$+\dfrac{15}{52}$

6. _____

Objective 3   Subtract unlike fractions.

For extra help, see Example 4 on pages 217–218 of your text and Section Lecture video for Section 3.3 and Exercise Solutions Clip 27.

Subtract the following fractions. Simplify all answers.

7. $\dfrac{7}{8}-\dfrac{1}{2}$

7. _____

8. $\dfrac{3}{5}-\dfrac{1}{4}$

8. _____

9. $\dfrac{9}{10}-\dfrac{4}{25}$

9. _____

Name: Date:

Instructor: Section:

Objective 4   Subtract unlike fractions vertically.

For extra help, see Example 5 on page 218 of your text and Section Lecture video for Section 3.3 and Exercise Solutions Clip 33

Subtract the following fractions. Simplify all answers.

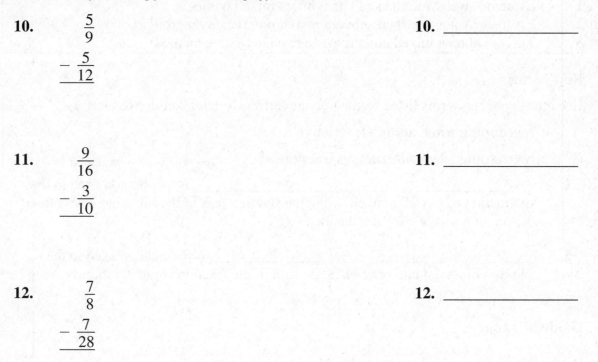

10. $\dfrac{5}{9}$

$-\dfrac{5}{12}$

10. _____

11. $\dfrac{9}{16}$

$-\dfrac{3}{10}$

11. _____

12. $\dfrac{7}{8}$

$-\dfrac{7}{28}$

12. _____

Chapter 3 ADDING AND SUBTRACTING FRACTIONS

3.4   Adding and Subtracting Mixed Numbers

Learning Objectives
1   Estimate an answer, then add or subtract mixed numbers.
2   Estimate an answer, then subtract mixed numbers by regrouping.
3   Add or subtract mixed numbers using an alternative method.

Key Terms

Use the vocabulary terms listed below to complete each statement in exercises 1–2.

regrouping when adding fractions

regrouping when subtracting fractions

1.   _____ is the method used in the subtraction of mixed numbers when the fraction part of the minuend is less than the fraction part of the subtrahend.

2.   _____ is the method used in the addition of mixed numbers when the sum of the fraction is greater than 1.

Guided Examples

Review these examples for Objective 1:

1.   First estimate the answer. Then add or subtract to find the exact answer.

    a. $15\frac{2}{7} + 4\frac{4}{7}$

    Estimate: *Exact:*

    15 ←__Rounds to__ $\left\{\ 15\frac{2}{7}\right.$

    $\underline{+\ 5}$ ←__Rounds to__ $\left\{+\ 4\frac{4}{7}\right.$

    20 $19\frac{6}{7}$

The exact answer $19\frac{6}{7}$ is reasonable because it is close to the estimate of 20.

Now Try:

1.   First estimate the answer. Then add or subtract to find the exact answer.

    a. $8\frac{5}{6} + 3\frac{2}{3}$

    _____

b. $9\frac{5}{6} - 2\frac{1}{9}$

Estimate: *Exact:*

$$10 \xleftarrow{\text{Rounds to}} \left\{ \quad 9\frac{5}{6} = \quad 9\frac{15}{18} \right.$$

$$\underline{-2} \xleftarrow{\text{Rounds to}} \left\{ -2\frac{1}{9} = -2\frac{2}{18} \right.$$

$$8 \qquad\qquad\qquad\qquad 7\frac{13}{18}$$

The exact answer $7\frac{13}{18}$ is reasonable because it is close to the estimate of 8.

2. First estimate, and then add $16\frac{4}{7} + 5\frac{6}{7}$.

Estimate: *Exact:*

$$17 \xleftarrow{\text{Rounds to}} \left\{ \quad 16\frac{4}{7} \right.$$

$$\underline{+6} \xleftarrow{\text{Rounds to}} \left\{ +5\frac{6}{7} \right.$$

$$23 \qquad\qquad\qquad 21\frac{10}{7}$$

Because the improper fraction $\frac{10}{7}$ can be written as $1\frac{3}{7}$, the simplified sum is

$$21\frac{10}{7} = 21 + \frac{10}{7} = 21 + 1\frac{3}{7} = 22\frac{3}{7}$$

The estimate was 23, so the exact answer of $22\frac{3}{7}$ is reasonable.

Review these examples for Objective 2:

3. First estimate, and then subtract to find the exact answer.

a. $8 - 3\frac{7}{8}$

Estimate: *Exact:*

$$8 \xleftarrow{\text{Rounds to}} \left\{ \quad 8 \right.$$

$$\underline{-4} \xleftarrow{\text{Rounds to}} \left\{ -3\frac{7}{8} \right.$$

$$4$$

It is not possible to subtract $\frac{7}{8}$ without

b. $20\frac{3}{8} - 15\frac{1}{4}$

_____

2. First estimate, and then add $8\frac{2}{9} + 12\frac{8}{9}$.

_____

Now Try:

3. First estimate, and then subtract to find the exact answer.

a. $16 - 7\frac{5}{9}$

_____

regrouping the whole number 8 first.

$$8 = 7 + 1 = 7 + \frac{8}{8} = 7\frac{8}{8}$$

Now you can subtract.

$$
\begin{array}{r}
8 = 7\frac{8}{8} \\
-3\frac{7}{8} = -3\frac{7}{8} \\
\hline
4\frac{1}{8}
\end{array}
$$

The estimate was 4, so the exact answer of $4\frac{1}{8}$ is reasonable.

b. $9\frac{1}{4} - 5\frac{2}{3}$

Estimate: *Exact:*

$$
\begin{array}{ll}
9 \xleftarrow{\text{ Rounds to }} \left\{ \begin{array}{l} 9\frac{1}{4} = 9\frac{3}{12} \\ \\ \end{array} \right. \\
\underline{-6} \xleftarrow{\text{ Rounds to }} \left\{ \begin{array}{l} -5\frac{2}{3} = -5\frac{8}{12} \\ \hline \end{array} \right. \\
3
\end{array}
$$

It is not possible to subtract $\frac{3}{12}$ from $\frac{8}{12}$, so regroup the whole number 9.

$$9\frac{3}{12} = 9 + \frac{3}{12} = 8 + 1 + \frac{3}{12}$$

$$= 8 + \frac{12}{12} + \frac{3}{12}$$

$$= 8\frac{15}{12}$$

Now you can subtract.

$$
\begin{array}{r}
9\frac{1}{4} = 9\frac{3}{12} = 8\frac{15}{12} \\
-5\frac{8}{12} = -5\frac{8}{12} = -5\frac{8}{12} \\
\hline
3\frac{7}{12}
\end{array}
$$

The exact answer is $3\frac{7}{12}$, which is reasonable because it is close to the estimate of 3.

b. $8\frac{1}{6} - 4\frac{7}{12}$

_____

Name: Date:
Instructor: Section:

Review these examples for Objective 3: | **Now Try:**
4.   Add or subtract. | 4.   Add or subtract.

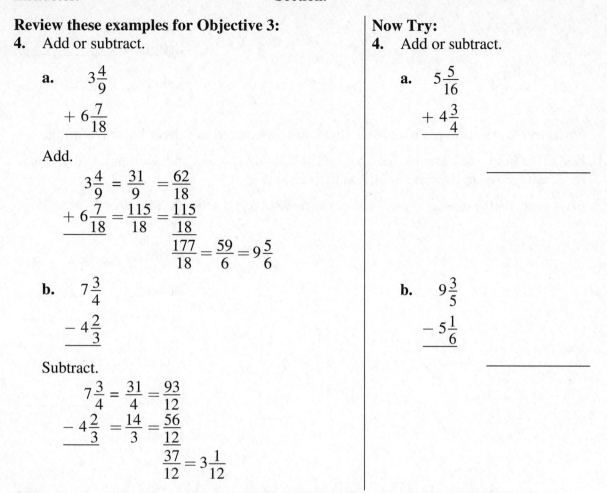

a. $3\frac{4}{9}$

$+\ 6\frac{7}{18}$

Add.

$$3\frac{4}{9} = \frac{31}{9} = \frac{62}{18}$$
$$+\ 6\frac{7}{18} = \frac{115}{18} = \frac{115}{18}$$
$$\frac{177}{18} = \frac{59}{6} = 9\frac{5}{6}$$

b. $7\frac{3}{4}$

$-\ 4\frac{2}{3}$

Subtract.

$$7\frac{3}{4} = \frac{31}{4} = \frac{93}{12}$$
$$-\ 4\frac{2}{3} = \frac{14}{3} = \frac{56}{12}$$
$$\frac{37}{12} = 3\frac{1}{12}$$

a. $5\frac{5}{16}$

$+\ 4\frac{3}{4}$

_____

b. $9\frac{3}{5}$

$-\ 5\frac{1}{6}$

_____

Objective 1   Estimate an answer, then add or subtract mixed numbers.

For extra help, see Examples 1–2 on pages 223-224 of your text and Section Lecture video for Section 3.4 and Exercise Solutions Clip 19.

First estimate the answer. Then add or subtract to find the exact answer. Write answers as mixed numbers in lowest terms.

1. $3\frac{1}{9}$

$+\ 4\frac{7}{8}$

1.
Estimate_____

Exact _____

2. $126\frac{4}{5}$

$28\frac{9}{10}$

$+\ 13\frac{2}{15}$

2.
Estimate_____

Exact _____

3. $14\dfrac{4}{7}$

     $-\,8\dfrac{1}{8}$

3.

Estimate_____

Exact _____

Objective 2   Estimate an answer, then subtract mixed numbers by regrouping.

For extra help, see Example 3 on pages 224-226 of your text and Section Lecture video for Section 3.4 and Exercise Solutions Clip 25 and 33.

First estimate the answer. Then subtract to find the exact answer. Simplify all answers

4. $11\dfrac{1}{4}$

     $-\,6\dfrac{3}{4}$

4.

Estimate_____

Exact _____

5. $129\dfrac{2}{3}$

     $-\,98\dfrac{14}{15}$

5.

Estimate_____

Exact _____

6. 42

     $-\,19\dfrac{3}{4}$

6.

Estimate_____

Exact _____

Objective 3   Add or subtract mixed numbers using an alternative method.

For extra help, see Example 4 on page 226 of your text and Section Lecture video for Section 3.4 and Exercise Solutions Clip 45 and 53.

Add or subtract by changing mixed numbers to improper fractions. Simplify all answers.

7. $5\dfrac{1}{3}$

     $+\,2\dfrac{5}{6}$

7. _____

Name: Date:
Instructor: Section:

8. $3\dfrac{1}{2}$

       $-1\dfrac{2}{3}$

       _____

8. _____

9. $9\dfrac{1}{8}$

       $-7\dfrac{4}{9}$

       _____

9. _____

125

Chapter 3 ADDING AND SUBTRACTING FRACTIONS

3.5   Order Relations and the Order of Operations

Learning Objectives
1   Identify the greater or lesser of two fractions.
2   Use exponents with fractions.
3   Use the order of operations with fractions.

Key Terms

Use the vocabulary terms listed below to complete each statement in exercises 1–2.

>   >   <

1.   The symbol _____ means " is less than."

2.   The symbol _____ means " is greater than."

Guided Examples

Review these examples for Objective 1:

1.   Rewrite the following using < and > symbols.

    a. $\dfrac{1}{7}$ is less than $\dfrac{9}{5}$

$\dfrac{1}{7}$ is less than $\dfrac{9}{5}$ is written as $\dfrac{1}{7} < \dfrac{9}{5}$.

    b. $\dfrac{10}{7}$ is greater than 1

$\dfrac{10}{7}$ is greater than 1 is written as $\dfrac{10}{7} > 1$.

    c. $\dfrac{7}{4}$ is less than $\dfrac{15}{7}$

$\dfrac{7}{4}$ is less than $\dfrac{15}{7}$ is written as $\dfrac{7}{4} < \dfrac{15}{7}$.

2.   Determine which fraction in each pair is greater.

    a. $\dfrac{11}{12}, \dfrac{14}{15}$

First, write the fractions as like fractions. The least common multiple for 12 and 15 is 60.

$$\dfrac{11}{12} = \dfrac{11 \cdot 5}{12 \cdot 5} = \dfrac{55}{60} \quad \text{and} \quad \dfrac{14}{15} = \dfrac{14 \cdot 4}{15 \cdot 4} = \dfrac{56}{60}$$

Now Try:

1.   Rewrite the following using < and > symbols.

    a. $\dfrac{1}{8}$ is less than $\dfrac{8}{3}$

_____

    b. $\dfrac{13}{4}$ is greater than 1

_____

    c. $\dfrac{9}{2}$ is less than $\dfrac{21}{4}$

_____

2.   Determine which fraction in each pair is greater.

    a. $\dfrac{5}{6}, \dfrac{7}{9}$

_____

Look at the numerators. Because 56 is greater than 55, $\frac{56}{60}$ is greater than $\frac{55}{60}$. Then because $\frac{56}{60}$ is equivalent to $\frac{14}{15}$,

$$\frac{14}{15} > \frac{11}{12} \quad \text{or} \quad \frac{11}{12} < \frac{14}{15}.$$

The greater fraction is $\frac{14}{15}$.

b. $\frac{7}{6}, \frac{19}{18}$

The least common multiple of 6 and 18 is 18.

$$\frac{7}{6} = \frac{7 \cdot 3}{6 \cdot 3} = \frac{21}{18} \quad \text{and} \quad \frac{19}{18} = \frac{19}{18}$$

This shows that $\frac{7}{6}$ is greater than $\frac{19}{18}$, or

$$\frac{7}{6} > \frac{19}{18}.$$

b. $\frac{20}{9}, \frac{21}{8}$

_____

Review these examples for Objective 2:

3. Simplify.

a. $\left(\frac{1}{3}\right)^3$

Three factors of $\frac{1}{3}$

$$\left(\frac{1}{3}\right)^3 = \frac{1}{3} \cdot \frac{1}{3} \cdot \frac{1}{3} = \frac{1}{27}$$

b. $\left(\frac{4}{11}\right)^2$

Two factors of $\frac{4}{11}$

$$\left(\frac{4}{11}\right)^2 = \frac{4}{11} \cdot \frac{4}{11} = \frac{16}{121}$$

Now Try:

3. Simplify.

a. $\left(\frac{1}{4}\right)^3$

_____

b. $\left(\frac{6}{7}\right)^2$

_____

c. $\left(\dfrac{5}{9}\right)^2 \cdot \left(\dfrac{3}{5}\right)^3$

Two factors of $\dfrac{5}{9}$, three factors of $\dfrac{3}{5}$

$$\left(\dfrac{5}{9}\right)^2 \cdot \left(\dfrac{3}{5}\right)^3 = \left(\dfrac{5}{9} \cdot \dfrac{5}{9}\right) \cdot \left(\dfrac{3}{5} \cdot \dfrac{3}{5} \cdot \dfrac{3}{5}\right)$$

$$= \dfrac{\overset{1}{\cancel{5}} \cdot \overset{1}{\cancel{5}} \cdot \overset{1}{\cancel{3}} \cdot \overset{1}{\cancel{3}} \cdot \overset{1}{\cancel{3}}}{\underset{1}{\underset{\cancel{3}}{\cancel{9}}} \cdot \underset{3}{\cancel{9}} \cdot \underset{1}{\cancel{5}} \cdot \underset{1}{\cancel{5}} \cdot 5}$$

$$= \dfrac{1}{15}$$

c. $\left(\dfrac{3}{4}\right)^2 \cdot \left(\dfrac{4}{5}\right)^2$

_____

Review these examples for Objective 3:

4. Simplify by using the order of operations.

a. $\dfrac{1}{4} + \dfrac{1}{3}\left(\dfrac{6}{5}\right)$

Multiply $\dfrac{1}{3}\left(\dfrac{6}{5}\right)$ first because multiplication and division are done before adding.

$$\dfrac{1}{4} + \dfrac{1}{\cancel{3}}\left(\dfrac{\overset{2}{\cancel{6}}}{5}\right) = \dfrac{1}{4} + \dfrac{2}{5}$$

Now, add. The least common denominator of 4 and 5 is 20.

$$\dfrac{1}{4} + \dfrac{2}{5} = \dfrac{5}{20} + \dfrac{8}{20} = \dfrac{13}{20}$$

b. $\dfrac{5}{6}\left(\dfrac{1}{4} + \dfrac{1}{5}\right)$

Work inside parentheses first.

$$\dfrac{5}{6}\left(\dfrac{1}{4} + \dfrac{1}{5}\right) = \dfrac{5}{6}\left(\dfrac{5}{20} + \dfrac{4}{20}\right)$$

$$= \dfrac{5}{6}\left(\dfrac{9}{20}\right)$$

$$= \dfrac{\overset{1}{\cancel{5}}}{\underset{2}{\cancel{6}}}\left(\dfrac{\overset{3}{\cancel{9}}}{\underset{4}{\cancel{20}}}\right)$$

$$= \dfrac{3}{8}$$

Now Try:

4. Simplify by using the order of operations.

a. $\dfrac{7}{9}\left(\dfrac{4}{7} \cdot \dfrac{5}{6}\right)$

_____

b. $\dfrac{7}{6}\left(\dfrac{3}{5}\right) - \left(\dfrac{4}{5}\right)^2$

_____

c. $\left(\dfrac{3}{4}\right)^2 - \dfrac{4}{9}\left(\dfrac{3}{5}\right)$

Simplify the expression with the exponent.

$$\left(\dfrac{3}{4}\right)^2 - \dfrac{4}{9}\left(\dfrac{3}{5}\right) = \dfrac{9}{16} - \dfrac{4}{9}\left(\dfrac{3}{5}\right)$$

$$= \dfrac{9}{16} - \dfrac{4}{\overset{}{\underset{3}{\cancel{9}}}}\left(\dfrac{\overset{1}{\cancel{3}}}{5}\right)$$

$$= \dfrac{9}{16} - \dfrac{4}{15}$$

$$= \dfrac{135}{240} - \dfrac{64}{240}$$

$$= \dfrac{71}{240}$$

c. $\dfrac{\left(\dfrac{3}{4}\right)^2}{\left(\dfrac{5}{6}\right)}$

_____

Objective 1   Identify the greater or lesser of two factions.

For extra help, see Examples 1–2 on pages 239-240 of your text and Section Lecture video for Section 3.5 and Exercise Solutions Clip 13.

Write > or < to make a true statement.

1. $\dfrac{3}{8} \underline{\hspace{1cm}} \dfrac{5}{16}$ 1. _____

2. $\dfrac{23}{40} \underline{\hspace{1cm}} \dfrac{17}{30}$ 2. _____

3. $\dfrac{13}{24} \underline{\hspace{1cm}} \dfrac{23}{36}$ 3. _____

Objective 2   Use exponents with fractions.

For extra help, see Example 3 on page 241 of your text and Section Lecture video for Section 3.5 and Exercise Solutions Clip 25, 31, and 39

Simplify. Write the answer in lowest terms.

4. $\left(\dfrac{1}{2}\right)^2$ 4. _____

5. $\left(\dfrac{3}{2}\right)^4$ 5. _____

6. $\left(\dfrac{12}{7}\right)^2$ 6. _____

Objective 3   Use the order of operations with fractions.

For extra help, see Example 4 on page 242 of your text and Section Lecture video for Section 3.5 and Exercise Solutions Clip 63.

Simplify. Write the answer in lowest terms.

7. $\left(\dfrac{4}{5}\right)^2 \cdot \dfrac{5}{12}$ 7. _____

8. $\dfrac{1}{2} \cdot \dfrac{4}{5} + \dfrac{2}{3} \cdot \dfrac{9}{5}$ 8. _____

9. $\left(\dfrac{8}{7} - \dfrac{9}{14}\right) \div \dfrac{3}{7}$ 9. _____

Chapter 4 DECIMALS

4.1   Reading and Writing Decimals

Learning Objectives
1   Write parts of a whole using decimals.
2   Identify the place value of a digit.
3   Read and write decimals in words.
4   Write decimals as fractions or mixed numbers.

Key Terms

Use the vocabulary terms listed below to complete each statement in exercises 1–3.

      decimals   **decimal point**   **place value**

1.   We use _____ to show parts of a whole.

2.   A _____ is assigned to each place to the left or right of the decimal point.

3.   The dot that separates the whole number part from the fractional part of a decimal number is called the _____.

Guided Examples

Review these examples for Objective 1:
1.   Given a fraction and how it is read, write as a decimal.

    a. $\dfrac{7}{10}$ seven tenths

Decimal: 0.7

    b. $\dfrac{3}{100}$ three hundredths

Decimal: 0.03

    c. $\dfrac{63}{100}$ sixty-three hundredths

Decimal: 0.63

    d. $\dfrac{9}{1000}$ nine thousandths

Decimal: 0.009

Now Try:
1.   Given a fraction and how it is read, write as a decimal.

    a. $\dfrac{6}{10}$ six tenths

_____

    b. $\dfrac{5}{100}$ five hundredths

_____

    c. $\dfrac{37}{100}$ thirty-seven hundredths

_____

    d. $\dfrac{7}{1000}$ seven thousandths

_____

e. $\dfrac{56}{1000}$ fifty-six thousandths

Decimal: 0.056

f. $\dfrac{943}{1000}$ nine hundred forty-three thousandths

Decimal: 0.943

e. $\dfrac{49}{1000}$ forty-nine thousandths

_____

f. $\dfrac{518}{1000}$ five hundred eighteen thousandths

_____

Review these examples for Objective 2:
2. Identify the place value of each digit.

    a. 486.92

    4 hundreds
    8 tens
    6 ones
    .
    9 tenths
    2 hundredths

    b. 0.00465

    0 ones
    .
    0 tenths
    0 hundredths
    4 thousandths
    6 ten-thousandths
    5 hundred-thousandths

Now Try:
2. Identify the place value of each digit.

    a. 862.93

    _____

    b. 0.00769

    _____

Review these examples for Objective 3:
3. Tell how to read each decimal in words.

    a. 0.7

    Read it as: seven tenths

    b. 0.62

    Read it as: sixty-two hundredths

    c. 0.09

    Read it as: nine hundredths

    d. 0.352

    Read it as: three hundred fifty-two thousandths

    e. 0.0205

    Read it as: two hundred five ten-thousandths

Now Try:
3. Tell how to read each decimal in words.

    a. 0.9

    _____

    b. 0.53

    _____

    c. 0.07

    _____

    d. 0.502

    _____

    e. 0.0469

    _____

4. Read each decimal.

  a. 17.8

  seventeen and eight tenths

  b. 543.71

  five hundred forty-three and seventy-one hundredths

  c. 0.089

  eighty-nine thousandths

  d. 12.4053

  twelve and four thousand fifty-three ten-thousandths

4. Read each decimal.

  a. 4.7

  _____

  b. 18.009

  _____

  c. 0.0082

  _____

  d. 57.906

  _____

Review these examples for Objective 4:

5. Write each decimal as a fraction or mixed number.

  a. 0.16

  The digits to the right of the decimal point, 16, are the numerator of the fraction. The denominator is 100 for hundredths because the rightmost digit is in the hundredths place.

$$0.16 = \frac{16}{100}$$

  b. 0.739

  The rightmost digit is in the thousandths place.

$$0.739 = \frac{739}{1000}$$

  c. 5.0088

  The whole number part stays the same. The rightmost digit is in the ten-thousandths place.

$$5.0088 = 5\frac{88}{10,000}$$

6. Write each decimal as a fraction or mixed number in lowest terms.

  a. 0.6

$$0.6 = \frac{6}{10} \qquad \text{Write } \frac{6}{10} \text{ in lowest terms.}$$

$$\frac{6}{10} = \frac{6 \div 2}{10 \div 2} = \frac{3}{5}$$

Now Try:

5. Write each decimal as a fraction or mixed number.

  a. 0.27

  _____

  b. 0.303

  _____

  c. 3.2636

  _____

6. Write each decimal as a fraction or mixed number in lowest terms.

  a. 0.2

  _____

b. 0.88

$$0.88 = \frac{88}{100} = \frac{88 \div 4}{100 \div 4} = \frac{22}{25}$$

c. 17.216

$$17.216 = 17\frac{216}{1000} = 17\frac{216 \div 8}{1000 \div 8} = 17\frac{27}{125}$$

d. 53.6065

$$53.6065 = 53\frac{6065}{10,000} = 53\frac{6065 \div 5}{10,000 \div 5} = 53\frac{1213}{2000}$$

b. 0.35

_____

c. 6.04

_____

d. 562.0404

_____

Objective 1   Write parts of a whole using decimals.

For extra help, see Example 1 on page 265 of your text and Section Lecture video for Section 4.1.

Write the portion of each square that is shaded as a fraction, as a decimal, and in words.

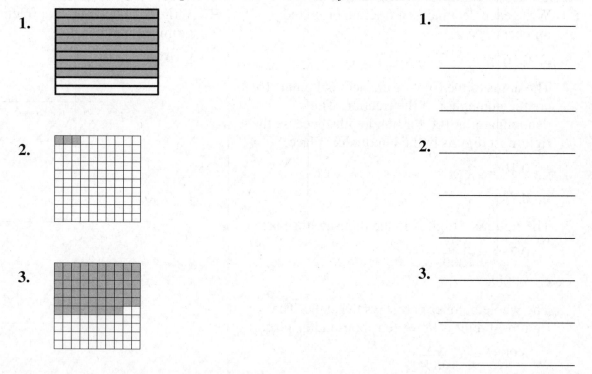

1.

1. _____

_____

_____

2.

2. _____

_____

_____

3.

3. _____

_____

_____

Objective 2   Identify the place value of a digit.

For extra help, see Example 2 on page 266 of your text and Section Lecture video for Section 4.1 and Exercise Solutions Clip 5 and 11.

Identify the digit that has the given place value.

4.   43.507   tenths

4. _____

                    hundredths

_____

5.   2.83714   thousandths   5. _____

                     ten-thousandths   _____

Identify the place value of each digit in these decimals.

6.   37.082   3   6. _____

                      7   _____

                      0   _____

                      8   _____

                      2   _____

Objective 3   Read and write decimals in words.

For extra help, see Examples 3–4 on pages 266–267 of your text and Section Lecture video for Section 4.1 and Exercise Solutions Clip 39 and 43.

Tell how to read each decimal in words.

7.   0.08   7. _____

8.   10.835   8. _____

Write each decimal in numbers

9.   Thirty eight and fifty-two hundred thousandths   9. _____

Objective 4   Write decimals as fractions or mixed numbers.

For extra help, see Examples 5–6 on page 268 of your text and Section Lecture video for Section 4.1 and Exercise Solutions Clip 21 and 31.

Write each decimal as a fraction or mixed number in lowest terms.

10.   0.001   10. _____

11.   3.6   11. _____

12.   0.95   12. _____

Chapter 4 DECIMALS

4.2   Rounding Decimal Numbers

Learning Objectives
1   Learn the rules for rounding decimals.
2   Round decimals to any given place.
3   Round money amounts to the nearest cent or nearest dollar.

Key Terms

Use the vocabulary terms listed below to complete each statement in exercises 1–2.

       rounding   **decimal places**

1.   _____ are the number of digits to the right of the decimal point.

2.   When we "cut off" a number after a certain place value, we are _____ that number.

Guided Examples

Review these examples for Objective 2:

1.   Round 16.87453 to the nearest thousandth.

Step 1 Draw a "cut-off" line after the thousandths place.
   16.874 | 53

Step 2 Look only at the first digit you are cutting off. Ignore the other digits you are cutting off.
   16.874 | 53   (Ignore the 3)

Step 3 If the first digit you are cutting off is 5 or more, round up the part of the number you are keeping.
    16.874 | 53
  + 0.001
    16.875

So, 16.87453 rounded to the nearest thousandth is 16.875. We can write $16.87453 \approx 16.875$.

2.   Round to the place indicated.

a. 6.4387 to the nearest tenth

Step 1 Draw a cut-off line after the tenths place.
   6.4 | 387

Step 2 Look only at the 3.
   6.4 | 387   (Ignore the 8 and 7)

Now Try:

1.   Round 43.80290 to the nearest thousandth.

          _____

2.   Round to the place indicated.

a. 0.7976 to the nearest hundredth

          _____

Step 3 The first digit is 4 or less, so the part you are keeping stays the same.

   6.4 | 387
   ―――
   6.4

Rounding 6.4387 to the nearest tenth is 6.4. We can write $6.4387 \approx 6.4$.

b. 0.79846 to the nearest hundredth

Step 1 Draw a cut-off line after the hundredths place.

   0.79 | 846

Step 2 Look only at the 8.

   0.79 | 846

Step 3 The first digit is 5 or more, so round up by adding 1 hundredth to the part you are keeping.

      1
   0.79 | 846
 + 0.01
  ――――
   0.80

0.79846 rounded to the nearest hundredth is 0.80. We can write $0.79846 \approx 0.80$.

c. 0.02709 to the nearest thousandth

   0.027 | 09

The first digit cut is 4 or less, so the part you are keeping stays the same.
0.02709 rounded to the nearest thousandth is 0.027. We can write $0.02709 \approx 0.027$.

d. 64.983 to the nearest tenth

   64.9 | 83

The first digit cut is 5 or more, so round up by adding 1 tenth to the part you are keeping.

     1
   64.9 | 83
 + 0.1
  ――――
   65.0

64.983 rounded to the nearest tenth is 65.0. We can write $64.983 \approx 65.0$. You must write the 0 in the tenths place to show that the number was rounded to the nearest tenth.

b. 7.7804 to the nearest hundredth

_____

c. 22.0397 to the nearest thousandth

_____

d. 0.649 to the nearest tenth

_____

Review these examples for Objective 3:

3.   Round each money amount to the nearest cent.

   a. $6.5348

   Is $6.5348 closer to $6.53 or to $6.54?
   The first digit cut is 4 or less, so the part you are keeping stays the same.
   $6.53 | 48
   You pay $6.53.

   b. $0.895

   Is $0.895 closer to $0.89 or $0.90?
   $0.89 | 5
   The first digit cut is 5 or more, so round up.

   $$\begin{array}{r} \overset{1}{\$0.89}\ |\ 5 \\ +\ \$0.01 \\ \hline \$0.90 \end{array}$$
   You pay $0.90.

4.   Round to the nearest dollar.

   a. $56.76

   Draw a cut-off line after the ones place.
   $56|.76
   First digit cut is 5 or more, so round up by adding $1.

   $$\begin{array}{r} \$56|.76 \\ +\quad 1 \\ \hline \$57 \end{array}$$
   So, $56.76 rounded to the nearest dollar is $57.

   b. $697.41

   Draw a cut-off line after the ones place.
   $697|.41
   First digit cut is 4 or less, so the part you keep stays the same.
   So, $697.41 rounded to the nearest dollar is $697.

   c. $599.66

   Draw a cut-off line after the ones place.
   $599|.66
   First digit cut is 5 or more, so round up by adding $1.

Now Try:

3.   Round each money amount to the nearest cent.

   a. $2.0849

   _____

   b. $425.0954

   _____

4.   Round to the nearest dollar.

   a. $37.81

   _____

   b. $307.20

   _____

   c. $880.83

   _____

$599|.66

+ _____1

$600

So, $599.66 rounded to the nearest dollar is $600.

d. $5779.50

$5779|.50

First digit cut is 5 or more, so round up by adding $1.

$5779|.50

+ _____1

$5780

$5779.50 rounded to the nearest dollar is $5780.

e. $0.53

$0|.53

First digit cut is 5 or more, so round up.
$0.53 rounded to the nearest dollar is $1.

d. $6859.77

_____

e. $0.61

_____

Objective 1   Learn the rules for rounding decimals.

For extra help, see page 274 of your text and Section Lecture video for Section 4.2.

Select the phrase that makes the sentence correct.

1. When rounding a number to the nearest tenth, if the   1. _____
    digit in the hundredths place is 5 or more, round the
    digit in the tenths place (up/down).

2. When rounding a number to the nearest hundredth,   2. _____
    look at the digit in the (tenth/thousandth) place.

Objective 2   Round decimals to any given place.

For extra help, see Examples 1–2 on pages 274–276 of your text and Section Lecture video for Section 4.2 and Exercise Solutions Clip 9, 11, 15, and 19.

Round each number to the place indicated.

3. 489.84 to the nearest tenth   3. _____

4. 54.4029 to the nearest hundredth   4. _____

5. 989.98982 to the nearest thousandth   5. _____

Objective 3   Round money amounts to the nearest cent or nearest dollar.

For extra help, see Examples 3–4 on page 277–278 of your text and Section Lecture video for Section 4.2 and Exercise Solutions Clip 21 and 23.

Round to the nearest dollar.

   6. $28.39   **6.** _____

   7. $11,839.73   **7.** _____

Round to the nearest cent.

   8. $1028.6666   **8.** _____

Chapter 4 DECIMALS

4.3   Adding and Subtracting Decimal Numbers

Learning Objectives
1   Add decimals.
2   Subtract decimals.
3   Estimate the answer when adding or subtracting decimals.

Key Terms

Use the vocabulary terms listed below to complete each statement in exercises 1–2.

      estimating   **front end rounding**

1.   With _____, we round to the highest possible place.

2.   Avoid common errors in working decimal problems by _____
     the answer first.

Guided Examples

Review these examples for Objective 1:

1.   Find each sum.

    a. 29.73 and 56.84

Step 1 Write the numbers in columns with the decimal points lined up.

$$\begin{array}{r} 29.73 \\ +56.84 \end{array}$$

Step 2 Add as if these were whole numbers.
Step 3 Line up decimal point in answer under the decimal points in problem.

$$\begin{array}{r} {\scriptstyle 1\ 1} \\ 29.73 \\ +56.84 \\ \hline 86.57 \end{array}$$

    b. 8.437 + 5.361 + 13.295

Write the numbers vertically with decimal points lined up. Then add.

$$\begin{array}{r} {\scriptstyle 1\ 1\ \ 1\ 1} \\ 8.437 \\ 5.361 \\ +13.295 \\ \hline 27.093 \end{array}$$

Now Try:

1.   Find each sum.

    a. 12.687 + 2.943

    _____

    b. 0.428 + 16.005 + 5.276

    _____

2. Find each sum.

  a. $6.7 + 0.41$

  There are two decimal places in 0.41, so write a 0 in the hundredths place in 6.7 so it has two decimal places also.

$$\begin{array}{r} 6.70 \\ +0.41 \\ \hline 7.11 \end{array}$$

  b. $12 + 9.36 + 3.754$

  Write in zeros so that all the addends have three decimal places.

$$\begin{array}{r} 12.000 \\ 9.360 \\ +\ \ 3.754 \\ \hline 25.114 \end{array}$$

2. Find each sum.

  a. $7.53 + 29.314$

  _____

  b. $0.631 + 999.3 + 14$

  _____

Review these examples for Objective 2:

3. Find each difference.

  a. 14.32 from 36.74

  Step 1 Line up decimal points.

$$\begin{array}{r} 36.74 \\ -14.32 \end{array}$$

  Step 2 Both numbers have two decimal places; no need to write in zeros.

  Step 3 Line up decimal point in answer.

$$\begin{array}{r} 36.74 \\ -14.32 \\ \hline 22.42 \end{array}$$

  b. 167.53 minus 69.85

  Regrouping is needed here.

$$\begin{array}{r} {\scriptstyle 0\ \ 15\ \ 16\ \ \ 14\ \ 13} \\ \not{1}\,\not{6}\,\not{7}.\not{5}\,\not{3} \\ -\ \ \ \ 69.85 \\ \hline 97.68 \end{array}$$

  Line up decimal point in answer.

Now Try:

3. Find each difference.

  a. 7.352 from 18.964

  _____

  b. 50.43 minus 39.86

  _____

4. Find each difference.

   a. 19.6 from 38.264

Line up decimal points and write in zeros so both numbers have three decimal places.

$$
\begin{array}{r}
38.264 \\
-19.600 \\
\hline
18.664
\end{array}
$$

   b. 28.8 – 19.963

Write in two zeros and subtract as usual.

$$
\begin{array}{r}
28.800 \\
-19.963 \\
\hline
8.837
\end{array}
$$

   c. 16 less 7.54

Write a decimal point and two zeros after 16. Subtract as usual.

$$
\begin{array}{r}
16.00 \\
-\ \ 7.54 \\
\hline
8.46
\end{array}
$$

4. Find each difference.

   a. 3.87 from 8.524

       _____

   b. 20 – 16.74

       _____

   c. 1 less 0.499

       _____

Review these examples for Objective 3:

5. Use front end rounding to round each number. Then add or subtract the rounded numbers to get an estimate answer. Finally, find the exact answer.

   a. Find the sum of 295.8 and 7.894.

Estimate: *Exact:*

$$
\begin{array}{r}
300 \\
+\ 8 \\
\hline
308
\end{array}
\quad
\begin{array}{l}
\xleftarrow{\text{Rounds to}} \\
\xleftarrow{\text{Rounds to}}
\end{array}
\quad
\begin{array}{r}
295.800 \\
+\ 7.894 \\
\hline
303.694
\end{array}
$$

The estimate goes out to the hundreds place and so does the exact answer. Therefore, the decimal point is probably in the correct place in the exact answer.

   b. $79.53 + $24.35

Estimate: *Exact:*

$$
\begin{array}{r}
\$80 \\
+\ 20 \\
\hline
\$100
\end{array}
\quad
\begin{array}{l}
\xleftarrow{\text{Rounds to}} \\
\xleftarrow{\text{Rounds to}}
\end{array}
\quad
\begin{array}{r}
\$79.53 \\
+\ 24.35 \\
\hline
\$103.88
\end{array}
$$

Exact answer is close to estimate, so it is reasonable.

Now Try:

5. Use front end rounding to round each number. Then add or subtract the rounded numbers to get an estimate answer. Finally, find the exact answer.

   a. Find the sum of 5.74 and 8.107.

       _____

   b. $39.43 + $3.52

       _____

c. Find the difference between 0.89 m and 7 m.

Use subtraction to find the difference between the two numbers. The larger number, 7, is written on top.

Estimate: *Exact:*

$$7 \xleftarrow{\text{Rounds to}} 7.00$$
$$\underline{-1} \xleftarrow{\text{Rounds to}} \underline{-0.89}$$
$$6 \qquad\qquad 6.11 \text{ m}$$

d. Subtract 3.7482 from 9.4.

Estimate: *Exact:*

$$9 \xleftarrow{\text{Rounds to}} 9.4000$$
$$\underline{-4} \xleftarrow{\text{Rounds to}} \underline{-3.7482}$$
$$5 \qquad\qquad 5.6518$$

c. Find the difference between 16.945 ft and 36.087 ft.

_____

d. Subtract 13.846 from 49.

_____

Objective 1   Add decimals.

For extra help, see Examples 1–2 on pages 281–282 of your text and Section Lecture video for Section 4.3 and Exercise Solutions Clip 5 and 9.

Find each sum.

1.   43.96 + 48.53

1. _____

2.   87.6 + 90.4

2. _____

3.   45.83 + 20.923 + 5.7

3. _____

Objective 2   Subtract decimals.

For extra help, see Examples 3–4 on pages 282–283 of your text and Section Lecture video for Section 4.3 and Exercise Solutions Clip 13.

Find each difference.

4.   84.6 – 18.1

4. _____

5. $69.524 - 26.958$ **5.** _____

6. $71 - 12.68$ **6.** _____

Objective 3   Estimate the answer when adding or subtracting decimals.

For extra help, see Example 5 on page 284 of your text and Section Lecture video for Section 4.3 and Exercise Solutions Clip 25.

First, use front end rounding and estimate each answer. Then add or subtract to find the exact answer.

7.   593.8
         27.93
      $+ \; 54.87$ **7.**
                                          Estimate_____

                                          Exact _____

8.   20.85
      $- \;\; 7.69$ **8.**
                                          Estimate_____

                                          Exact _____

9.   9.7
      $- \; 4.862$ **9.**
                                          Estimate_____

                                          Exact _____

Chapter 4 DECIMALS

4.4   Multiplying Decimal Numbers

Learning Objectives
1   Multiply decimals.
2   Estimate the answer when multiplying decimals.

Key Terms

Use the vocabulary terms listed below to complete each statement in exercises 1–3.

decimal places   factor   product

1.   Each number in a multiplication problem is called a _____.

2.   When multiplying decimal numbers, first multiply the numbers, then find the total number of _____ in both factors.

3.   The answer to a multiplication problem is called the _____.

Guided Examples

Review this example for Objective 1:

1.   Find the product of 6.23 and 5.4.

Step 1 Multiply the numbers as if they were whole numbers.

$$\begin{array}{r} 6.\,2\,3 \\ \times\quad 5.\,4 \\ \hline 2\,4\,9\,2 \\ 3\,1\,1\,5 \\ \hline 3\,3\,6\,4\,2 \end{array}$$

Step 2 Count the total number of decimal places in both factors.

$$\begin{array}{r} 6.\,2\,3 \;\leftarrow 2 \text{ decimal places} \\ \times\quad 5.\,4 \;\leftarrow 1 \text{ decimal place} \\ \hline 2\,4\,9\,2 \quad 3 \text{ total decimal places} \\ 3\,1\,1\,5 \\ \hline 3\,3\,6\,4\,2 \end{array}$$

Step 3 Count over 3 places in the product and write the decimal point. Count from right to left.

Now Try:

1.   Find the product of 2.51 and 4.3.

_____

$$
\begin{array}{r}
6.\,2\,3 \leftarrow 2 \text{ decimal places} \\
\times\quad 5.\,4 \leftarrow 1 \text{ decimal place} \\
\hline
2\,4\,9\,2 \qquad 3 \text{ total decimal places} \\
3\,1\,1\,5 \\
\hline
3\,3.\,6\,4\,2
\end{array}
$$

2. Find the product: $(0.035)(0.07)$.

Start by multiply, then count decimal places.

$$
\begin{array}{r}
0.\,0\,3\,5 \leftarrow 3 \text{ decimal places} \\
\times\quad 0.\,0\,7 \leftarrow 2 \text{ decimal places} \\
\hline
2\,4\,5 \qquad 5 \text{ total decimal places}
\end{array}
$$

After multiplying, the answer has only three decimal places, but five are needed, so write two zeros on the left side of the answer. Then count over 5 places and write in the decimal point.

$$
\begin{array}{r}
0.\,0\,3\,5 \leftarrow 3 \text{ decimal places} \\
\times\quad 0.\,0\,7 \leftarrow 2 \text{ decimal places} \\
\hline
.0\,0\,2\,4\,5 \qquad 5 \text{ total decimal places}
\end{array}
$$

The final product is 0.00245, which has five decimal places.

2. Find the product $(0.062)(0.03)$.

_____

Review this example for Objective 2:

3. First estimate the answer to $(74.56)(18.9)$ using front end rounding. Then find the exact answer.

Estimate:

$$
\begin{array}{r}
70 \leftarrow \text{Rounds to} \\
\times\,20 \leftarrow \text{Rounds to} \\
\hline
1400
\end{array}
$$

Exact:

$$
\begin{array}{r}
7\,4.\,5\,6 \leftarrow 2 \text{ decimal places} \\
\times\quad 1\,8.9 \leftarrow 1 \text{ decimal place} \\
\hline
6\,7\,1\,0\,4 \qquad 3 \text{ total decimal places} \\
5\,9\,6\,4\,8 \\
7\,4\,5\,6 \\
\hline
1\,4\,0\,9.\,1\,8\,4
\end{array}
$$

Both the estimate and the exact answer go out to the thousands place, so the decimal point in 1409.184 is probably in the correct place.

Now Try:

3. First estimate the answer to $(22.43)(5.03)$ using front end rounding. Then find the exact answer.

_____

Objective 1   Multiply decimals.

For extra help, see Examples 1–2 on pages 289–290 of your text and Section Lecture video for Section 4.4 and Exercise Solutions Clip 5, 15, and 19.

Find each product.

1.
$$
\begin{array}{r}
19.3 \\
\times\,4.7 \\
\hline
\end{array}
$$

1. _____

2.   0.682   **2.** _____
      × 3.9

3.   (0.074)(0.05)   **3.** _____

Objective 2   Estimate the answer when multiplying decimals.

For extra help, see Example 3 on page 290 of your text and Section Lecture video for Section 4.4 and Exercise Solutions Clip 23.

First use front-end rounding and estimate the answer. Then multiply to find the exact answer.

4.   29.8   **4.**
      × 3.4   **Estimate**_____

                                     Exact _____

5.   32.53   **5.**
      × 23.26   **Estimate**_____

                                     Exact _____

6.   391.9   **6.**
      × 7.74   **Estimate**_____

                                     Exact _____

Chapter 4 DECIMALS

4.5   Dividing Decimal Numbers

Learning Objectives
1   Divide a decimal by a whole number.
2   Divide a number by a decimal.
3   Estimate the answer when dividing decimals.
4   Use the order of operations with decimals.

Key Terms

Use the vocabulary terms listed below to complete each statement in exercises 1–4.

        repeating decimal   **quotient**   **dividend**   **divisor**

1.   In a division problem, the number being divided is called the _____.

2.   The number $0.8\bar{3}$ is an example of a _____.

3.   The answer to a division problem is called the _____.

4.   In the problem $6.39 \div 0.9$, 0.9 is called the _____.

Guided Examples

Review these examples for Objective 1:

1.   Find each quotient. Check the quotients by multiplying.

    a. 24.48 by 4

Rewrite the division problem.
Step 1 Write the decimal point in the quotient directly above the decimal point in the dividend.

$$\begin{array}{r} . \\ 4\overline{)24.48} \end{array}$$

Step 2 Divide as if the numbers were whole numbers.

$$\begin{array}{r} 6.12 \\ 4\overline{)24.48} \end{array}$$

Check by multiplying the quotient times the divisor.

$$\begin{array}{r} 6.12 \\ \times 4 \\ \hline 24.48 \end{array}$$

The quotient (answer) is 6.12.

Now Try:

1.   Find each quotient. Check the quotients by multiplying.

    a. 9.891 by 7

_____

b. $8\overline{)425.6}$ **b.** $5\overline{)75.15}$

Write the decimal point in the quotient above the
decimal point in the dividend. Then divide as if
the number were whole numbers.

$$
\begin{array}{r}
53.2 \\
8\overline{)425.6} \\
\underline{40} \\
25 \\
\underline{24} \\
16 \\
\underline{16} \\
0
\end{array}
\qquad
\text{Check:}
\begin{array}{r}
53.2 \\
\times\ \ 8 \\
\hline
425.6
\end{array}
$$

The quotient is 53.2.

_____

2. Divide 1.35 by 4. Check the quotient by   **2.** Divide 1008.9 by 50. Check the
multiplying.   quotient by multiplying.

Divide.

$$
\begin{array}{r}
0.33 \\
4\overline{)1.35} \\
\underline{1\,2} \\
15 \\
\underline{12} \\
3
\end{array}
$$

_____

Write a 0 after the 5 in the dividend so you can
continue dividing. Keep writing more zeros in
the dividend, if needed.

$$
\begin{array}{r}
0.3375 \\
4\overline{)1.3500} \\
\underline{1\,2} \\
15 \\
\underline{12} \\
30 \\
\underline{28} \\
20 \\
\underline{20} \\
0
\end{array}
\qquad
\text{Check:}
\begin{array}{r}
0.3375 \\
\times\ \ 4 \\
\hline
1.3500
\end{array}
$$

The quotient is 0.3375.

3. Divide 8.87 by 9. Round the quotient to the nearest thousandth.

Write extra zeros in the dividend so you can continue dividing.

$$
\begin{array}{r}
0.9855 \\
9\overline{)8.8700} \\
\underline{8\ 1} \\
77 \\
\underline{72} \\
50 \\
\underline{45} \\
50 \\
\underline{45} \\
5
\end{array}
$$

Notice that the digit 5 in the answer is repeating. There are two ways to show that the answer is a repeating decimal that goes on forever.

    $0.9855\ldots$ or $0.98\overline{5}$

To round to thousandths, divide out one more place, to ten-thousandths.

    $8.87 \div 9 = 0.9855\ldots$ rounds to 0.986.

Check the answer by multiplying 0.986 by 9. Because 0.986 is a rounded answer, the check will not give exactly 8.87, but it should be very close.

    $(0.986)(9) = 8.874$

3. Divide 302.24 by 18. Round the quotient to the nearest thousandth.

_____

Review these examples for Objective 2:

4.

   a. $0.005\overline{)41.2}$

Move the decimal point in the divisor three places to the right so 0.005 becomes the whole number 5. Move the decimal point in the dividend the same number of places and write in two extra 0s.

$$
\begin{array}{r}
8240. \\
5\overline{)41200.}
\end{array}
$$

Now Try:

4.

   a. $0.0024\overline{)48.984}$

_____

b. Divide 7 by 2.4. Round to the nearest hundredth.

Move the decimal point in the divisor one place to the right so 2.4 becomes the whole number 24. The decimal point in the dividend starts on the right side of 7 and is also moved one place to the right.
(Remember, in order to round to hundredths, divide out one more place, to thousandths.)

```
        2.916
   24) 70.000
        48
        220
        216
         40
         24
        160
        144
         16
```

Round the quotient. It is 2.92.

b. Divide 8 by 4.5. Round to the nearest hundredth.

_____

Review this example for Objective 3:

5. First use front end rounding to round each number and estimate the answer. Then divide to find the exact answer.

$$767.38 \div 3.7$$

Estimate:
```
      200
   4) 800
```

Exact:
```
        27.4
   37) 7673.8
        74
        273
        259
        148
        148
          0
```

Notice that the estimate, which is in hundreds, is very different from the exact answer, which is in tens.

Find the error and rework.
The exact answer is 207.4, which fits the estimate of 200.

Now Try:

5. First use front end rounding to round each number and estimate the answer. Then divide to find the exact answer.

$$185.22 \div 4.9$$

_____

Review these examples for Objective 4:

6. Use the order of operations to simplify each expression.

    a. $4.5 + 8.2^2 + 11.63$

    Apply the exponent.
      $4.5 + 67.24 + 11.63$
    Add from left to right.
      $71.74 + 11.63$
      83.37

    b. $1.73 + (3.8 - 2.9)(6.5)$

    Work inside the parentheses.
      $1.73 + (0.9)(6.5)$
    Multiply.
      $1.73 + 5.85$
    Add.
      7.58

    c. $5.6^2 - 1.4 \div 7(2.4)$

    Apply the exponent.
      $31.36 - 1.4 \div 7(2.4)$
    Multiply and divide from left to right.
      $31.36 - 0.2(2.4)$
    Multiply.
      $31.36 - 0.48$
    Subtract last.
      30.88

Now Try:

6. Use the order of operations to simplify each expression.

    a. $5.9 - 0.48 + 2.6^2$

    _____

    b. $4.06 \div 1.4 \times 7.8$

    _____

    c. $0.07 + 0.3(8 - 6.99)$

    _____

Objective 1   Divide a decimal by a whole number.

For extra help, see Examples 1–3 on pages 297–299 of your text and Section Lecture video for Section 4.5 and Exercise Solutions Clip 5 and 7.

Find each quotient. Round answers to the nearest thousandth, if necessary.

1. $5)\overline{34.8}$

1. _____

2. $11)\overline{46.98}$

2. _____

3. $33\overline{)77.847}$

3. _____

Objective 2   Divide a number by a decimal.

For extra help, see Example 4 on page 300 of your text and Section Lecture video for Section 4.5.

Find each quotient. Round answers to the nearest thousandth, if necessary.

4. $0.9\overline{)3.4166}$

4. _____

5. $3.4\overline{)436.05}$

5. _____

6. $0.07 \div 0.00043$

6. _____

Objective 3   Estimate the answer when dividing decimals.

For extra help, see Example 5 on page 301 of your text and Section Lecture video for Section 4.5 and Exercise Solutions Clip 25.

*Decide if each answer is **reasonable** or **unreasonable** by rounding the numbers and estimating the answer.*

7. $126.2 \div 11.2 = 11.268$

7. _____

8. $31.5 \div 8.4 = 37.5$

8. _____

9. $8695.15 \div 98.762 = 88.0415$

9. _____

Name: Date:
Instructor: Section:

Objective 4   Use the order of operations with decimals.

For extra help, see Example 6 on pages 301–302 of your text and Section Lecture video for Section 4.5 and Exercise Solutions Clip 53 and 55.

Use the order of operations to simplify each expression.

10. $3.1^2 - 1.9 + 5.8$

10. _____

11. $58.1 - (17.9 - 15.2) \times 1.8$

11. _____

12. $9.1 - 0.07(2.1 \div 0.042)$

12. _____

Chapter 4 DECIMALS

4.6   Fractions as Decimals

Learning Objectives
1   Write fractions as equivalent decimals.
2   Compare the size of fractions and decimals.

Key Terms

Use the vocabulary terms listed below to complete each statement in exercises 1–4.

numerator   denominator   mixed number   equivalent

1.   A fraction and a decimal that represent the same portion of a whole are

_____.

2.   The _____ of a fraction is the dividend.

3.   The _____ of a fraction shows the number of equal parts in a
whole.

4.   A _____ consists of a whole number part and a fractional or
decimal part.

Guided Examples

Review these examples for Objective 1:
1.   Write the fraction as a decimal.

   a. $\dfrac{7}{8}$

   $\dfrac{7}{8}$ means $7 \div 8$. Write it as $8\overline{)7}$. Write extra
   zeros in the dividend so you can continue
   dividing until the remainder is zero.

$$
\begin{array}{r}
0.875 \\
8\overline{)7.000} \\
\underline{6\,4} \\
60 \\
\underline{56} \\
40 \\
\underline{40} \\
0
\end{array}
$$

   Therefore, $\dfrac{7}{8} = 0.875$

Now Try:
1.   Write the fraction as a decimal.

   a. $\dfrac{1}{16}$

   _____

b. $3\dfrac{3}{16}$

One method is to divide 3 by 16 to get 0.1875 for the fraction part. Then add the whole number part to 0.1875.

$$\dfrac{3}{16} \rightarrow 16\overline{)\begin{array}{l}0.1875\\3.0000\end{array}} \rightarrow \begin{array}{r}3.0000\\+\ 0.1875\\\hline 3.1875\end{array}$$

$$\begin{array}{r}16\\\hline 140\\128\\\hline 120\\112\\\hline 80\\80\\\hline 0\end{array}$$

So $3\dfrac{3}{16} = 3.187$.

A second method is to first write $3\dfrac{3}{16}$ as an improper fraction and then divide numerator by denominator.

$$3\dfrac{3}{16} = \dfrac{51}{16}$$

$$\dfrac{51}{16} \rightarrow 51 \div 16 \rightarrow 16\overline{)51} \rightarrow 16\overline{)\begin{array}{l}3.1875\\51.0000\end{array}}$$

$$\begin{array}{r}48\\\hline 30\\16\\\hline 140\\128\\\hline 120\\112\\\hline 80\\80\\\hline 0\end{array}$$

So $3\dfrac{3}{16} = 3.1875$.

2. Write $\dfrac{5}{9}$ as a decimal and round to the nearest thousandth.

$\dfrac{5}{9}$ means $5 \div 9$. To round to thousandths, divide out one more place, to ten-thousandths.

b. $4\dfrac{3}{8}$

_____

2. Write $\dfrac{5}{12}$ as a decimal and round to the nearest thousandth.

_____

$$\frac{5}{9} \rightarrow 5 \div 9 \rightarrow 9\overline{)5} \rightarrow 9\overline{)5.0000}$$

$$\begin{array}{r} 0.5555 \\ \underline{45} \\ 50 \\ \underline{45} \\ 50 \\ \underline{45} \\ 50 \\ \underline{45} \\ 5 \end{array}$$

Written as a repeating decimal, $\frac{5}{9} = 0.\overline{5}$.

Rounded to the nearest thousandth, $\frac{5}{9} = 0.556$.

Review these examples for Objective 2:	**Now Try:**
3. Write in <, >, or = in the blank between each pair of numbers.	**3.** Write in <, >, or = in the blank between each pair of numbers.
a. 0.735 _____ 0.75	**a.** 0.85 _____ 0.7895
Because 0.735 is to the left of 0.75, use the < symbol. 0.735 is less than 0.75 can be written as 0.735 < 0.75	_____
b. $\dfrac{3}{5}$ _____ 0.6	**b.** $\dfrac{5}{16}$ _____ 0.3125
$\dfrac{3}{5}$ and 0.6 are the same point on the number line. They are equivalent. $\dfrac{3}{5} = 0.6$	_____
c $0.\overline{3}$ _____ 0.3	**c** $\dfrac{3}{7}$ _____ $0.\overline{4}$

0.3 is to the right of $0.\overline{3}$ (which is actually 0.333...), so use the > symbol.
    $0.\overline{3}$ is greater than 0.3 can be written as
    $0.\overline{3} > 0.3$
	_____

d $\dfrac{14}{18}$ _____ $0.\overline{7}$	**d** $\dfrac{27}{33}$ _____ $0.\overline{81}$
Write $\dfrac{14}{18}$ in lowest terms as $\dfrac{7}{9}$. We see that $\dfrac{7}{9} = 0.\overline{7}$.	_____

4. Write each group of numbers in order, from least to greatest.

   a. 0.58   0.575   0.5816

   Write zeros to the right of 0.58 and 0.575, so they also have four decimal places. Then find the least and greatest number of ten-thousandths.

   $0.58 = 0.5800 = 5800$ ten-thousandths middle

  $0.575 = 0.5750 = 5750$ ten-thousandths least

        $= 0.5816 = 5816$ ten-thousandths greatest

   From least the greatest, the correct order is

     0.575   0.58   0.5816

   b. $4\frac{7}{8}$ 4.7   4.82

   Write $4\frac{7}{8}$ as $\frac{39}{8}$ and divide to get the decimal form 4.875. Then because 4.875 has three decimal places, write zeros so all the numbers have three decimal places.

    $4\frac{7}{8} = 4.875 = 4$ and 875 thousandths greatest

    $4.7 = 4.700 = 4$ and 700 thousandths least

    $4.82 = 4.820 = 4$ and 820 thousandths middle

   From least to greatest, the correct order is

    4.7   4.82 $4\frac{7}{8}$

4. Write each group of numbers in order, from least to greatest.

   a. 0.3   0.307   0.3057

             _____

   b. $\frac{2}{9}$, $\frac{3}{13}$, 0.23, $\frac{1}{5}$

             _____

Objective 1   Write fractions as equivalent decimals.

For extra help, see Examples 1–2 on pages 307–308 of your text and Section Lecture video for Section 4.6 and Exercise Solutions Clip 9, 23, and 28.

Write each fraction or mixed number as a decimal. Round to the nearest thousandth, if necessary.

1. $\frac{1}{8}$

1. _____

2. $4\frac{1}{9}$

2. _____

3. $19\frac{17}{24}$

3. _____

Name: _____   Date: _____

Instructor: _____   Section: _____

Objective 2   Compare the size of fractions and decimals.

For extra help, see Examples 3–4 on pages 309–310 of your text and Section Lecture video for Section 4.6 and Exercise Solutions Clip 53 and 55.

Write < or > to make a true statement.

4. $\dfrac{5}{6}$ ____ 0.83   4. _____

Arrange in order from smallest to largest.

5. $\dfrac{3}{11}, \dfrac{1}{3}, 0.29$ 5. _____

6. $1.085, 1\dfrac{5}{11}, 1\dfrac{7}{20}$ 6. _____

5.   11 to $2\frac{4}{9}$

5. _____

Solve. Write each ratio as a fraction in lowest terms.

6.   One car has a $15\frac{1}{2}$ gallon gas tank while another has a 22 gallon gas tank. Find the ratio of the amount the first tank holds to the amount the second tank holds.

6. _____

Objective 3   Solve ratio problems after converting units.

For extra help, see Example 5 on pages 333–334 of your text and Section Lecture video for Section 5.1 and Exercise Solutions Clip 23 and 27.

Write each ratio as a fraction in lowest terms. Be sure to convert units as necessary.

7.   4 days to 2 weeks

7. _____

8.   6 yards to 10 feet

8. _____

9.   40 ounces to 3 pounds

9. _____

Chapter 5 RATIO AND PROPORTION

5.2   Rates

Learning Objectives
1   Write rates as fractions.
2   Find unit rates.
3   Find the best buy based on cost per unit.

Key Terms

Use the vocabulary terms listed below to complete each statement in exercises 1–3.

> **rate**   **unit rate**   **cost per unit**

1.   When the denominator of a rate is 1, it is called a _____.

2.   The _____ is that rate that tells how much is paid for one item.

3.   A _____ compares two measurements with different units.

Guided Examples

Review these examples for Objective 1:
1.   Write each rate as a fraction in lowest terms.

   a. 7 gallons for $56

   Write the units: gallons and dollars
   $$\frac{7 \text{ gallons} \div 7}{56 \text{ dollars} \div 7} = \frac{1 \text{ gallon}}{8 \text{ dollars}}$$

   b. 75 inches of growth in 15 weeks

   $$\frac{75 \text{ inches} \div 15}{15 \text{ weeks} \div 15} = \frac{5 \text{ inches}}{1 \text{ week}}$$

   c. 984 miles on 32 gallons of gas

   $$\frac{984 \text{ miles} \div 8}{32 \text{ gallons} \div 8} = \frac{123 \text{ miles}}{4 \text{ gallons}}$$

Now Try:
1.   Write each rate as a fraction in lowest terms.

   a. $7 for 35 pages

   _____

   b. 300 strokes in 20 minutes

   _____

   c. 396 strawberries for 24 cakes

   _____

Review these examples for Objective 2:

2.   Find each unit rate.

   a. 478.5 miles on 14.5 gallons of gas

Write the rate as a fraction.

$$\frac{478.5 \text{ miles}}{14.5 \text{ gallons}}$$

Divide 478.5 by 14.5 to find the unit rate.

$$145\overline{)4785} \quad \text{(quotient } 33\text{)}$$

$$\frac{478.5 \text{ miles} \div 14.5}{14.5 \text{ gallons} \div 14.5} = \frac{33 \text{ miles}}{1 \text{ gallon}}$$

The unit rate is 33 miles per gallon
or 33 miles/gallon.

   b. 770 miles in 22 hours

$$\frac{770 \text{ miles}}{22 \text{ hours}} \quad \text{Divide: } 22\overline{)770} \quad \text{(quotient } 35\text{)}$$

The unit rate is 35 miles/hour.

   c. $1240 in 8 days

$$\frac{1240 \text{ dollars}}{8 \text{ days}} \quad \text{Divide: } 8\overline{)1240} \quad \text{(quotient } 155\text{)}$$

The unit rate is $155/day.

Review these examples for Objective 3:

3.   Determine the best price for peanut butter. For
18 ounces the price is $2.89, for 28 ounces the
price is $3.99, and for 40 ounces, the price is
$6.18.

For 18 ounces, the cost per ounce is

$$\frac{\$2.89}{18 \text{ ounces}} \approx \$0.16$$

For 28 ounces, the cost per ounce is

$$\frac{\$3.99}{28 \text{ ounces}} \approx \$0.14$$

For 40 ounces, the cost per ounce is

$$\frac{\$6.18}{40 \text{ ounces}} \approx \$0.15$$

The lowest cost per ounce is $0.14, so
the 28-ounce container is the best buy.

Now Try:

2.   Find each unit rate.

   a. 294 miles on 10.5 gallons of
gas

_____

   b. $7.56 for 6 pounds of apples

_____

   c. $580 in 4 days

_____

Now Try:

3.   Find the best buy:
2 pints for $3.55,
3 pints for $5.25, and
5 pints for $8.50.

_____

Name: _____   Date: _____

Instructor: _____   Section: _____

4. Solve each application problem.

a. Brand AA laundry detergent costs $5.99 for 32 ounces. Brand ZZ laundry detergent costs $13.29 for 100 ounces. Which choice is the best buy?

To find Brand AA's unit cost, divide $5.99 by 32 ounces. Similarly, to find Brand ZZ's unit cost, divide $13.29 by 100 ounces.

Brand AA $\quad \dfrac{\$5.99}{32 \text{ ounces}} \approx 0.187$ per ounce

Brand ZZ $\quad \dfrac{\$13.29}{100 \text{ ounces}} = 0.1329$ per ounce

Brand ZZ has the lower cost per ounce and is the better buy.

b. Special offers affect the best buy. Brand M of multivitamin costs $3.99 for 200 + 100 tablets (a bonus of 100 tablets). Brand N of multivitamin costs $5.47 for 200 tablets with a special Buy one Get one Free. Which choice is the best buy?

Brand M is $3.99 for 300 tablets (200 + 100)
Brand N is $5.47 for 400 tablets (200 + 200)

To find the best buy, divide to find the lowest cost per tablet.

Brand M $\quad \dfrac{\$3.99}{300 \text{ tablets}} \approx \$0.0133$ per tablet

Brand N $\quad \dfrac{\$5.47}{400 \text{ tablets}} \approx \$0.0137$ per tablet

Brand M has the lower cost per tablet and is the best buy.

4. Solve each application problem.

a. An eight-pack of AA-size batteries costs $4.99. A twenty-pack of AA-size batteries costs $12.99. Which battery pack is the best buy?

_____

b. Which shampoo is the best buy? You have a coupon for 75¢ off Brand S and a coupon for $1.25 off Brand T. Brand S is $4.79 for 14 ounces. Brand T is $5.69 for 18 ounces.

_____

Objective 1   Write rates as fractions.

For extra help, see Example 1 on page 339 of your text and Section Lecture video for Section 5.2 and Exercise Solutions Clip 3.

Write each rate as a fraction in lowest terms.

1.   119 pills for 17 patients

1. _____

2.   28 dresses for 4 women

2. _____

3.   256 pages for 8 chapters

3. _____

Objective 2   Find unit rates.

For extra help, see Example 2 on pages 340 of your text and Section Lecture video for Section 5.2 and Exercise Solutions Clip 11.

Find each unit rate.

4. $3500 in 20 days   4. _____

5. $7875 for 35 pounds   5. _____

6.   189.88 miles on 9.4 gallons   6. _____

Objective 3   Find the best buy based on cost per unit.

For extra help, see Examples 3–4 on pages 340–342 of your text and Section Lecture video for Section 5.2 and Exercise Solutions Clip 27 and 35.

Find the best buy (based on cost per unit) for each item.

7.   Peanut butter: 18 ounces for $1.77;   7. _____
     24 ounces for $2.08

8.   Batteries: 4 for $2.79; 10 for $4.19   8. _____

9.   Soup: 3 cans for $1.75; 5 cans for $2.75; 9. _____
     8 cans for $4.55

Chapter 5 RATIO AND PROPORTION

5.3   Proportions

Learning Objectives
1   Write proportions.
2   Determine whether proportions are true or false.
3   Find cross products.

Key Terms

Use the vocabulary terms listed below to complete each statement in exercises 1–2.

cross products   **proportion**

1.   A _____ shows that two ratios or rates are equivalent.

2.   To see whether a proportion is true, determine if the _____ are equal.

Guided Examples

Review these examples for Objective 1:
1.   Write each proportion.

    a. 7 m is to 13 m as 28 m is to 52 m

$$\frac{7 \text{ m}}{13 \text{ m}} = \frac{28 \text{ m}}{52 \text{ m}} \quad \text{so} \quad \frac{7}{13} = \frac{28}{52}$$

    b. $14 is to 8 gallons as $7 is to 4 gallons

$$\frac{\$14}{8 \text{ gallons}} = \frac{\$7}{4 \text{ gallons}}$$

Now Try:
1.   Write each proportion.

    a. 24 ft is to 17 ft as 72 ft is to 51 ft

    _____

    b. $10 is to 7 cans as $60 is to 42 cans

    _____

Review these examples for Objective 2:
2.   Determine whether each proportion is true or false by writing both ratios in lowest terms.

    a. $\dfrac{7}{11} = \dfrac{16}{24}$

Write each ratio in lowest terms.

$\dfrac{7}{11} \leftarrow$ Already in lowest terms $\dfrac{16 \div 8}{24 \div 8} = \dfrac{2}{3} \leftarrow$ Lowest terms

Because $\dfrac{7}{11}$ is not equivalent to $\dfrac{2}{3}$, the proportion is false.

Now Try:
2.   Determine whether each proportion is true or false by writing both ratios in lowest terms.

    a. $\dfrac{36}{28} = \dfrac{24}{18}$

    _____

b. $\dfrac{9}{15} = \dfrac{21}{35}$

Write each ratio in lowest terms.

$\dfrac{9 \div 3}{15 \div 3} = \dfrac{3}{5}$ $\dfrac{21 \div 7}{35 \div 7} = \dfrac{3}{5}$

Both ratios are equivalent to $\dfrac{3}{5}$, so the

proportion is true.

b. $\dfrac{4}{12} = \dfrac{9}{27}$

_____

Review these examples for Objective 3:

3. Use cross products to see whether each proportion is true or false.

a. $\dfrac{5}{8} = \dfrac{30}{48}$

Multiply along one diagonal, then multiply along the other diagonal.

$\dfrac{5}{8} = \dfrac{30}{48}$ $\nearrow 8 \cdot 30 = 240$
$\searrow 5 \cdot 48 = 240$

The cross products are equal, so the proportion is true.

b. $\dfrac{3\frac{1}{5}}{4\frac{2}{3}} = \dfrac{7}{10}$

$\nearrow 4\dfrac{2}{3} \cdot 7 = \dfrac{14}{3} \cdot \dfrac{7}{1} = \dfrac{98}{3} = 32\dfrac{2}{3}$

$\dfrac{3\frac{1}{5}}{4\frac{2}{3}} = \dfrac{7}{10}$

$\searrow 3\dfrac{1}{5} \cdot 10 = \dfrac{16}{5} \cdot \dfrac{\overset{2}{\cancel{10}}}{1} = \dfrac{32}{1} = 32$

The cross products are unequal, so the proportion is false.

Now Try:

3. Use cross products to see whether each proportion is true or false.

a. $\dfrac{6}{17} = \dfrac{18}{51}$

_____

b. $\dfrac{3.2}{5} = \dfrac{7}{10}$

_____

Objective 1   Write proportions.

For extra help, see Example 1 on pages 347 of your text and Section Lecture video for Section 5.3 and Exercise Solutions Clip 5.

Write each proportion.

1. 50 is to 8 as 75 is to 12.

1. _____

2. 36 is to 45 as 8 is to 10.

2. _____

3. 3 is to 33 as 12 is to 132.

3. _____

Objective 2   Determine whether proportions are true or false.

For extra help, see Example 2 on pages 347 of your text and Section Lecture video for Section 5.3 and Exercise Solutions Clip 11.

Determine whether each proportion is true or false by writing the ratios in lowest terms. Show the simplified ratios and then write **true** *or* **false**.

4. $\dfrac{48}{36} = \dfrac{3}{4}$

4. _____

5. $\dfrac{30}{25} = \dfrac{6}{5}$

5. _____

6. $\dfrac{63}{18} = \dfrac{56}{14}$

6. _____

Objective 3   Find cross products.

For extra help, see Example 3 on pages 348–349 of your text and Section Lecture video for Section 5.3 and Exercise Solutions Clip 23, 27, and 33.

Use cross products to determine whether each proportion is true or false. Show the cross products and then write **true** *or* **false**.

7. $\dfrac{28}{50} = \dfrac{49}{75}$

7. _____

8. $\dfrac{4\frac{3}{5}}{9} = \dfrac{18\frac{2}{5}}{36}$

8. _____

9. $\dfrac{2.98}{7.1} = \dfrac{1.7}{4.3}$

9. _____

Chapter 5 RATIO AND PROPORTION

5.4   Solving Proportions

Learning Objectives
1   Find the unknown number in a proportion.
2   Find the unknown number in a proportion with mixed numbers or decimals.

Key Terms

Use the vocabulary terms listed below to complete each statement in exercises 1–3.

cross products   proportion   ratio

1.   A _____ is a statement that two ratios are equal.

2.   The _____ of the proportion $\frac{a}{b} = \frac{c}{d}$ are ad and bc.

3.   A _____ is a comparison of two quantities with the same units.

Guided Examples

Review these examples for Objective 1:

1.   Find the unknown number in each proportion. Round answers to the nearest hundredth when necessary.

a. $\dfrac{14}{x} = \dfrac{21}{18}$

Recall that ratios can be rewritten in lowest terms. Write $\dfrac{21}{18}$ in lowest terms as $\dfrac{7}{6}$, which gives the proportion $\dfrac{14}{x} = \dfrac{7}{6}$.

Step 1 Find the cross product.

$$\dfrac{14}{x} = \dfrac{7}{6} \quad \begin{array}{l} \nearrow x \cdot 7 \\ \searrow 14 \cdot 6 \end{array}$$

Step 2 Show that cross products are equal.

$x \cdot 7 = 14 \cdot 6$

$x \cdot 7 = 84$

Step 3 Divide both sides by 7.

Now Try:

1.   Find the unknown number in each proportion. Round answers to the nearest hundredth when necessary.

a. $\dfrac{24}{x} = \dfrac{9}{12}$

_____

$$\frac{x \cdot \cancel{7}^{1}}{\cancel{7}_{1}} = \frac{84}{7}$$

$$x = 12$$

Step 4 Check in original proportion.

$$\frac{14}{12} = \frac{21}{18} \quad \nearrow 12 \cdot 21 = 252$$
$$\searrow 14 \cdot 18 = 252$$

The cross products are equal, so 12 is the correct solution.

b. $\frac{9}{13} = \frac{21}{x}$

b. $\frac{2}{3} = \frac{x}{16}$

_____

Step 1 Find the cross product.

$$\frac{9}{13} = \frac{21}{x} \quad \nearrow 13 \cdot 21 = 273$$
$$\searrow 9 \cdot x$$

Step 2 Show that cross products are equal.

$$9 \cdot x = 273$$

Step 3 Divide both sides by 9.

$$\frac{\cancel{9}^{1} \cdot x}{\cancel{9}_{1}} = \frac{273}{9}$$

$$x = 30.33 \text{ rounded to the nearest hundredth}$$

Step 4 Check in original proportion.

$$\frac{9}{13} = \frac{21}{30.33} \quad \nearrow 13 \cdot 21 = 273$$
$$\searrow 9 \cdot 30.33 = 272.97$$

The cross products are slightly different because of the rounded value of *x*. However, they are close enough to see that the problem was done correctly and that 30.33 is the approximate solution.

Review these examples for Objective 2:

2. Find the unknown number in each proportion.

a. $\frac{3\frac{1}{4}}{8} = \frac{x}{12}$

$$\frac{3\frac{1}{4}}{8} = \frac{x}{12} \quad \nearrow 8 \cdot x$$
$$\searrow 3\frac{1}{4} \cdot 12$$

Change $3\frac{1}{4}$ to an improper fraction and write in

Now Try:

2. Find the unknown number in each proportion.

a. $\frac{4\frac{1}{3}}{5} = \frac{x}{3}$

_____

lowest terms.

$$3\frac{1}{4} \cdot 12 = \frac{13}{4} \cdot \frac{12}{1} = \frac{13}{\cancel{4}} \cdot \frac{\cancel{12}^{3}}{1} = \frac{39}{1} = 39$$

Show that the cross products are equal.

$$8 \cdot x = 39$$

Divide both sides by 8.

$$\frac{\cancel{8}^{1} \cdot x}{\cancel{8}_{1}} = \frac{39}{8}$$

Write the solution as a mixed number in lowest terms.

$$x = \frac{39}{8} = 4\frac{7}{8}$$

The unknown number is $4\frac{7}{8}$.

Check in original proportion.

$$\nearrow 8 \cdot 4\frac{7}{8} = \frac{\cancel{8}^{1}}{1} \cdot \frac{39}{\cancel{8}_{1}} = 39$$

$$\frac{3\frac{1}{4}}{8} = \frac{4\frac{7}{8}}{12}$$

$$\searrow 3\frac{1}{4} \cdot 12 = \frac{13}{\cancel{4}_{1}} \cdot \frac{\cancel{12}^{3}}{1} = 39$$

The cross products are equal, so $4\frac{7}{8}$ is the correct solution.

b. $\dfrac{3.5}{1.4} = \dfrac{4}{x}$

Show that the products are equal.

$$(3.5)(x) = (1.4)(4)$$

$$(3.5)(x) = 5.6$$

Divide both sides by 3.5.

$$\frac{\cancel{(3.5)}^{1} \cdot (x)}{\cancel{3.5}_{1}} = \frac{5.6}{3.5}$$

$$x = \frac{5.6}{3.5}$$

Complete the division.

$$x = 1.6 \qquad 35\overline{)56.0}^{\,1.6}$$

b. $\dfrac{x}{8} = \dfrac{1.2}{1.5}$

_____

So, the unknown number is 1.6. Write the solution in the original proportion and check it by finding the cross products.

$$\frac{3.5}{1.4} = \frac{4}{1.6} \qquad \nearrow 1.4 \cdot 4 = 5.6 \\ \searrow 3.5 \cdot 1.6 = 5.6$$

The cross products are equal, so 1.6 is the correct solution.

Objective 1   Find the unknown number in a proportion.

For extra help, see Example 1 on pages 353–354 of your text and Section Lecture video for Section 5.4 and Exercise Solutions Clip 7, 9, and 17.

Find the unknown number in each proportion.

1. $\dfrac{9}{7} = \dfrac{x}{28}$ 1. _____

2. $\dfrac{7}{5} = \dfrac{98}{x}$ 2. _____

3. $\dfrac{100}{x} = \dfrac{75}{30}$ 3. _____

Objective 2   Find the unknown number in a proportion with mixed numbers or decimals.

For extra help, see Example 2 on pages 354–355 of your text and Section Lecture video for Section 5.4 and Exercise Solutions Clip 23.

Find the unknown number in each proportion. Write answers as a whole or a mixed number if possible.

4. $\dfrac{2}{3\frac{1}{4}} = \dfrac{8}{x}$ 4. _____

5. $\dfrac{3}{x} = \dfrac{0.8}{5.6}$ 5. _____

6. $\dfrac{2\frac{5}{9}}{x} = \dfrac{23}{\frac{3}{5}}$ 6. _____

Chapter 5 RATIO AND PROPORTION

5.5   Solving Application Problems with Proportions

Learning Objectives
1   Use proportions to solve application problems.

Key Terms

Use the vocabulary terms listed below to complete each statement in exercises 1–2.

      rate   **ratio**

1.   A statement that compares a number of inches to a number of inches is a
        _____ .

2.   A statement that compares a number of gallons to a number of miles is a
        _____ .

Guided Examples

Review these examples for Objective 1:

1.   Alexis drove 343 miles on 7.5 gallons of gas. How far can she travel on a full tank of 12 gallons of gas?

Step 1 Read the problem. The problem asks for the number of miles the car can travel on 12 gallons of gas.

Step 2 Work out a plan. Decide what is being compared and write a proportion using the two rates.

$$\frac{343 \text{ miles}}{7.5 \text{ gallons}} = \frac{x \text{ miles}}{12 \text{ gallons}}$$

Step 3 Estimate a reasonable answer. To estimate the answer, notice that 12 is a little more than 1.5 times 7.5. So use $343 \cdot 1.5 = 514.5$ miles, as an estimate.

Step 4 Solve the problem. Ignore the units while solving for x.

$$\frac{343 \text{ miles}}{7.5 \text{ gallons}} = \frac{x \text{ miles}}{12 \text{ gallons}}$$
$$(7.5)(x) = (343)(12)$$
$$(7.5)(x) = 4116$$

Now Try:

1.   Aiden spends 23 hours painting 4 apartments. How long will it take him to paint the other 16 apartments?

        _____

$$\frac{\cancel{(7.5)}^{1}(x)}{\cancel{7.5}_{1}} = \frac{4116}{7.5}$$

$$x = 548.8 \quad \text{Round to 549.}$$

Step 5 State the answer. Rounded to the nearest mile, the car can travel about 549 miles on a full tank of gas.

Step 6 Check your work. The answer 549 miles, is a little more than the estimate of 514.5 miles, so it is reasonable.

2.   There are 24 women in a college class of 39. At that rate, how many of the college's 10,400 students are women?

Step 1 Read the problem. The problem asks how many of the 10,400 students are women.

Step 2 Work out a plan. Decide what is being compared and write a proportion using the two rates.

$$\frac{24 \text{ women}}{39 \text{ students}} = \frac{x \text{ women}}{10,400 \text{ students}}$$

Step 3 Estimate a reasonable answer. To estimate the answer, notice that 24 is a little less than twice 39. Half of 10,400 is $10,400 \div 2 = 5200$, so our estimate is more than 5200 students.

Step 4 Solve the problem. Ignore the units while solving for x.

$$\frac{24 \text{ women}}{39 \text{ students}} = \frac{x \text{ women}}{10,400 \text{ students}}$$
$$39 \cdot x = 24 \cdot 10,400$$
$$39 \cdot x = 249,600$$
$$\frac{\cancel{39}^{1} \cdot x}{\cancel{39}_{1}} = \frac{249,600}{39}$$

$$x = 6400 \quad \text{No need to round.}$$

Step 5 State the answer. There are 6400 woman at the college.

Step 6 Check your work. The answer 6400 women, is a little more than the estimate of 5200 women, so it is reasonable.

2.   A survey showed that 4 out of 5 smokers have tried to quit smoking. At this rate, how many people in a group of 540 have tired to quit smoking?

_____

Name: _____   Date: _____

Instructor: _____   Section: _____

Objective 1   Use proportions to solve application problems.

For extra help, see Examples 1–2 on pages 360–362 of your text and Section Lecture video for Section 5.5 and Exercise Solutions Clip 3, 11, 17, and 23.

Set up and solve a proportion for each problem.

1. If 22 hats cost $198, find the cost of 12 hats.

   1. _____

2. If 150 square yards of carpet cost $3142.50, find the cost of 210 square yards of the carpet.

   2. _____

3. A biologist tags 50 deer and releases them in a wildlife preserve area. Over the course of a two-week period, she observes 80 deer, of which 12 are tagged. What is the estimate for the population of deer in this particular area?

   3. _____

Name: Date:
Instructor: Section:

Chapter 6 PERCENT

6.1   Basics of Percent

Learning Objectives
1   Learn the meaning of percent.
2   Write percents as decimals.
3   Write decimals as percents.
4   Understand 100%, 200%, and 300%.
5   Use 50%, 10%, and 1%.

Key Terms

Use the vocabulary terms listed below to complete each statement in exercises 1–3.

percent   **ratio**   **decimals**

1.   To compare two quantities that have the same type of units, use a _____.

2.   _____ means per one hundred.

3.   _____ represent parts of a whole.

Guided Examples

Review these examples for Objective 1:
1.   Find the percent.

   a. The tip is $15 for every $100 spent on food. What percent is this?

   The tip is $15 per $100, or $\frac{15}{100}$. The percent is 0.15.

   b. In a group of 100 students, 80 ride the bus. What percent ride the bus?

   Then 80 per 100, or $\frac{80}{100}$, or 80% of the students ride the bus.

Now Try:
1.   Find the percent.

   a. The sales tax is $9 per $100. What is the percent tax?

   _____

   b. In a group of 100 cars, 73 are a neutral color. What percent of the cars are a neutral color?

   _____

Review these examples for Objective 2:
2.   Write each percent as a decimal.

   a. 36%

   $36\% = 36 \div 100 = 0.36$

Now Try:
2.   Write each percent as a decimal.

   a. 29%

   _____

b. 14%

$14\% = 14 \div 100 = 0.14$

b. 86%

_____

3. Write each percent as a decimal by moving the decimal point two places to the left.

3. Write each percent as a decimal by moving the decimal point two places to the left.

a. 56%

a. 48%

Drop the percent sign and move the decimal point two places to the left.
$56\% = 56.\% = 0.56$

_____

b. 120%

b. 240%

$120\% = 120.\% = 1.20$ or 1.2

_____

c. 1.3%

c. 2.5%

0 is attached so the decimal point can be moved two places to the left.
$1.3\% = 0.013$

_____

d. 0.2%

d. 0.8%

Two zeros are attached so the decimal point can be moved two places to the left.
$0.2\% = 0.002$

_____

Review these examples for Objective 3:

4. Write each decimal as a percent by moving the decimal point two places to the right.

Now Try:

4. Write each decimal as a percent by moving the decimal point two places to the right.

a. 0.31

a. 0.43

Decimal point is moved two places to the right and percent symbol is attached.
$0.31 = 31\%$

_____

b. 1.6

b. 2.3

0 is attached so the decimal point can be moved two places to the right.
$1.6 = 1.60 = 160\%$

_____

c. 0.904

c. 0.751

$0.904 = 90.4\%$

_____

Review these examples for Objective 4:

5. Fill in the blanks.

Now Try:

5. Fill in the blanks.

a. 100% of $19 is _____ .

a. 100% of $35 is _____ .

100% is all of the money. So, 100% of $19 is ___$19___ .

_____

Name:
Date:
Instructor:
Section:

b. 200% of 170 miles is _____ .

200% is twice (2 times) as many miles.
So, 200% of 170 miles is __340 miles__ .

b. 200% of 86 miles is _____ .

_____

c. 300% of 76 plates is _____ .

300% is 3 times the number of plates.
So, 300% of 76 plates is __228 plates__ .

c. 300% of 52 glasses is _____ .

_____

Review these examples for Objective 5:
6. Fill in the blanks.

a. 10% of 300 years is _____ .

10% is $\frac{1}{10}$ of the years. Move the decimal point
one place to the left.
So, 10% of 300 years is __30 years__ .

b. 50% of 48 copies is _____ .

50% is half of the copies. So, 50% of 48 copies
is __24 copies__ .

c. 1% of 400 homes is _____ .

1% is $\frac{1}{100}$ of the homes. Move the decimal
point two places to the left.
So, 1% of 400 homes is __4 homes__ .

Now Try:
6. Fill in the blanks.

a. 10% of 200 students is

_____ .

_____

b. 50% of 54 photographs is

_____ .

_____

c. 1% of 500 miles is _____ .

_____

Objective 1 Learn the meaning of percent.

For extra help, see Example 1 on page 382 of your text and Section Lecture video for
Section 6.1

Write as a percent.

1. 68 people out of 100 drive small cars.

1. _____

2. The tax is $8 per $100.

2. _____

3. The cost for labor was $45 for every $100 spent to
manufacture an item.

3. _____

Objective 2   Write percents as decimals.

For extra help, see Examples 2–3 on pages 382–384 of your text and Section Lecture video for Section 6.1 and Exercise Solutions Clip 3, 11, 13, 15, and 17.

Write each percent as a decimal.

4.   42%   4. _____

5.   310%   5. _____

6.   18.9%   6. _____

Objective 3   Write decimals as percents.

For extra help, see Example 4 on pages 384–385 of your text and Section Lecture video for Section 6.1 and Exercise Solutions Clip 25.

Write each decimal as a percent.

7.   0.2   7. _____

8.   0.564   8. _____

9.   4.93   9. _____

Objective 4   Understand 100%, 200%, and 300%.

For extra help, see Example 5 on page 385 of your text and Section Lecture video for Section 6.1.

Fill in the blanks.

10.   100% of 12 dogs is _____.   10. _____

11.   200% of $520 is _____.   11. _____

12.   300% of $250 is _____.   12. _____

Objective 5   Use 50%, 10%, and 1%.

For extra help, see Example 6 on page 386 of your text and Section Lecture video for Section 6.1 and Exercise Solutions Clip 61.

Fill in the blanks.

13.   50% of 250 signs is _____   **13.** _____

14.   10% of 4920 televisions is _____.   **14.** _____

15.   1% of $98 is _____.   **15.** _____

Chapter 6 PERCENT

6.2   Percents and Fractions

Learning Objectives
1   Write percents as fractions.
2   Write fractions as percents.
3   Use the table of percent equivalents.

Key Terms

Use the vocabulary terms listed below to complete each statement in exercises 1–2.

percent   lowest terms

1.   A fraction is in _____ when its numerator and denominator have no common factor other than 1.

2.   A _____ can be written as a fraction with 100 in the denominator.

Guided Examples

Review these examples for Objective 1:

1.   Write each percent as a fraction or mixed number in lowest terms.

   a. 12%

   Recall writing 12% as a decimal.
   $$12\% = 12 \div 100 = 0.12$$
   Because 0.12 means 12 hundredths,
   $$0.12 = \frac{12}{100} = \frac{12 \div 4}{100 \div 4} = \frac{3}{25}$$

   b. 86%

   Write 86% as $\frac{86}{100}$.

   Write $\frac{86}{100}$ in lowest terms.

   $$\frac{86}{100} = \frac{86 \div 2}{100 \div 2} = \frac{43}{50}$$

   c. 350%

   $$350\% = \frac{350}{100} = \frac{350 \div 50}{100 \div 50} = \frac{7}{2} = 3\frac{1}{2}$$

Now Try:

1.   Write each percent as a fraction or mixed number in lowest terms.

   a. 30%

   _____

   b. 78%

   _____

   c. 175%

   _____

2. Write each percent as a fraction in lowest terms.

a. 62.5%

Write 62.5 over 100.
$$62.5\% = \frac{62.5}{100}$$
To get a whole number in the numerator, multiply the numerator and denominator by 10.
$$\frac{62.5}{100} = \frac{62.5(10)}{100(10)} = \frac{625}{1000}$$
Now write the fraction in lowest terms.
$$\frac{625}{1000} = \frac{625 \div 125}{1000 \div 125} = \frac{5}{8}$$

b. $16\frac{2}{3}\%$

Write $16\frac{2}{3}\%$ over 100.
$$16\frac{2}{3}\% = \frac{16\frac{2}{3}}{100}$$
When there is a mixed number in the numerator, rewrite the mixed number as an improper fraction.
$$\frac{16\frac{2}{3}}{100} = \frac{\frac{50}{3}}{100}$$
Next, rewrite the division problem in a horizontal form. Finally, multiply by the reciprocal of the divisor.
$$\frac{\frac{50}{3}}{100} = \frac{50}{3} \div 100 = \frac{50}{3} \div \frac{100}{1} = \frac{50}{3} \cdot \frac{1}{100} = \frac{1}{6}$$

2. Write each percent as a fraction in lowest terms.
a. 43.6%

_____

b. $22\frac{2}{9}\%$

_____

Review these examples for Objective 2:

3. Write each fraction as a percent. Round to the nearest tenth as necessary.

a. $\frac{7}{50}$

Write $\frac{7}{50}$ as a percent by solving for p in the proportion below.
$$\frac{7}{50} = \frac{p}{100}$$
Find cross products and show that they are equal.

Now Try:

3. Write each fraction as a percent. Round to the nearest tenth as necessary.

a. $\frac{7}{20}$

_____

$$50 \cdot p = 7 \cdot 100$$

$$50 \cdot p = 700$$

$$\frac{\overset{1}{\cancel{50}} \cdot p}{\underset{1}{\cancel{50}}} = \frac{700}{50}$$

$$p = 14$$

This result means that $\frac{7}{50} = \frac{14}{100}$ or 14%.

b. $\frac{37}{40}$

Write a proportion.

$$\frac{37}{40} = \frac{p}{100}$$

$$40 \cdot p = 37 \cdot 100$$

$$40 \cdot p = 3700$$

$$\frac{\overset{1}{\cancel{40}} \cdot p}{\underset{1}{\cancel{40}}} = \frac{3700}{40}$$

$$p = 92.5$$

So, $\frac{37}{40} = 92.5\%$.

c. $\frac{5}{9}$

Start with a proportion.

$$\frac{5}{9} = \frac{p}{100}$$

$$9 \cdot p = 5 \cdot 100$$

$$9 \cdot p = 500$$

$$\frac{\overset{1}{\cancel{9}} \cdot p}{\underset{1}{\cancel{9}}} = \frac{500}{9}$$

$$p = 55.\overline{5}$$

$$p \approx 55.6$$

So, $\frac{5}{9} = 55.\overline{5}\% \approx 55.6\%$.

b. $\frac{27}{40}$

c. $\frac{7}{18}$

Name: Date:
Instructor: Section:

Review these examples for Objective 3: | **Now Try:**

4. Find the following in the table in the text.   **4.** Find the following in the table from the text.

  a. $\frac{5}{8}$ as a percent   **a.** $\frac{1}{8}$ as a percent

  Find $\frac{5}{8}$ in the "fraction" column. The equivalent   _____
  percent is 62.5%.

  b. $0.\overline{3}$ as a fraction   **b.** 0.4 as a fraction

  Look in the "decimal" column for $0.\overline{3}$. The   _____
  equivalent fraction is $\frac{1}{3}$.

  c. $\frac{3}{5}$ as a percent   **c.** $\frac{1}{4}$ as a percent

  Find $\frac{3}{5}$ in the "fraction" column. The equivalent   _____
  percent is 60%.

Objective 1   Write percents as fractions.

For extra help, see Examples 1–2 on pages 393–394 of your text and Section Lecture video for Section 6.2 and Exercise Solutions Clip 5, 9, 11, and 17.

Write each percent as a fraction or mixed number in lowest terms.

  1. 140%   **1.** _____

  2. $18\frac{1}{3}\%$ **2.** _____

  3. 55.6%   **3.** _____

Objective 2   Write fractions as percents.

For extra help, see Example 3 on pages 395–396 of your text and Section Lecture video for Section 6.2 and Exercise Solutions Clip 31, 33, and 39.

Write each fraction or mixed number as a percent. Round percents to the nearest tenth, if necessary.

  4. $\frac{47}{50}$ **4.** _____

5. $\dfrac{11}{40}$ 5. _____

6. $\dfrac{64}{75}$ 6. _____

Objective 3   Use the table of percent equivalents.

For extra help, see Example 4 on page 396 of your text and Section Lecture video for Section 6.2 and Exercise Solutions Clip 49 and 51.

Complete this chart. Round decimals to the nearest thousandth and percents to the nearest tenth, if necessary.

Fraction	*Decimal*	*Percent*

7. $\dfrac{3}{8}$ _____   _____   7. _____

8.   _____   _____ $33\dfrac{1}{3}\%$ 8. _____

9.   _____   0.325   _____   9. _____

Chapter 6 PERCENT

**6.3   Using the Percent Proportion and Identifying the Components in a Percent
Problem**

Learning Objectives
1   Learn the percent proportion.
2   Solve for an unknown value in a percent proportion.
3   Identify the percent.
4   Identify the whole.
5   Identify the part.

Key Terms

Use the vocabulary terms listed below to complete each statement in exercises 1–3.

       percent proportion   **whole**   **part**

1.   The _____ in a percent problem is the entire
     quantity.

2.   The _____ in a percent problem is the portion being compared
     with the whole.

3.   Part is to whole as percent is to 100 is called the _____.

Guided Examples

Review these examples for Objective 1:
1.   Identify the component that is unknown: part,
whole, or percent. Do not solve the proportion.

    a. $\dfrac{5}{6} = \dfrac{\text{unknown}}{100}$

The unknown represents the percent component
of the percent proportion.

    b. $\dfrac{12}{\text{unknown}} = \dfrac{18}{100}$

The unknown represents the whole component
of the percent proportion.

    c. $\dfrac{\text{unknown}}{25} = \dfrac{36}{100}$

The unknown represents the part component of
the percent proportion.

Now Try:
1.   Identify the component that is
unknown: part, whole, or
percent. Do not solve the
proportion.

    a. $\dfrac{9}{\text{unknown}} = \dfrac{56}{100}$

        _____

    b. $\dfrac{\text{unknown}}{48} = \dfrac{18}{100}$

        _____

    c. $\dfrac{7}{9} = \dfrac{\text{unknown}}{100}$

        _____

Review these examples for Objective 2:

2. Use the percent proportion and solve for the unknown value. Let x represent the unknown value.

    a. part = 30, percent = 25; find the whole.

    Use the percent proportion, $\dfrac{\text{part}}{\text{whole}} = \dfrac{\text{percent}}{100}$.

    $\dfrac{30}{x} = \dfrac{25}{100}$ or $\dfrac{30}{x} = \dfrac{1}{4}$

    Find the cross products and show that the cross products are equal.

    $\dfrac{30}{x} = \dfrac{1}{4}$

    $x \cdot 1 = 30 \cdot 4$

    $x = 120$

    The whole is 120.

    b. part = 12, whole = 50

    Use the percent proportion.

    $\dfrac{12}{50} = \dfrac{x}{100}$

    $50 \cdot x = 12 \cdot 100$

    $50 \cdot x = 1200$

    $\dfrac{\overset{1}{\cancel{50}} \cdot x}{\underset{1}{\cancel{50}}} = \dfrac{1200}{50}$

    $x = 24$

    The percent is 24, written as 24%.

    c. whole = 48, percent = 25

    Use the percent proportion.

    $\dfrac{x}{48} = \dfrac{25}{100}$ or $\dfrac{x}{48} = \dfrac{1}{4}$

    $x \cdot 4 = 48 \cdot 1$

    $4 \cdot x = 48$

    $\dfrac{\overset{1}{\cancel{4}} \cdot x}{\underset{1}{\cancel{4}}} = \dfrac{48}{4}$

    $x = 12$

    The part is 12.

Now Try:

2. Use the percent proportion and solve for the unknown value. Let x represent the unknown value.

    a. part = 160, percent = 20; find the whole.

    _____

    b. part = 75, whole = 1500; find the percent.

    _____

    c. whole = 25, percent = 14; find the part.

    _____

Name: _____   Date: _____
Instructor: _____   Section: _____

Review these examples for Objective 3:	**Now Try:**
3. Find the percent in the following.	3. Find the percent in the following.
a. 43% of 1100 people exercise daily.	**a.** 950 tons of trash is 59% of what number of tons?
The percent is 43.	
	_____
b. $255 is 29 percent of what number?	**b.** Find the amount of sales tax by multiplying $650 and 7.75 percent.
The percent is 29, because 29 appears with the word percent.	
	_____
c. What percent of 6500 tons is 275 tons?	**c.** What percent of 2650 children attend preschool?
The word percent has no number with it, so the percent is the unknown part of the problem.	
	_____

Review these examples for Objective 4:	**Now Try:**
4. Identify the whole in the following.	4. Identify the whole in the following.
a. 43% of 1100 people exercise daily.	**a.** Find the amount of sales tax by multiplying $650 and 7.75 percent.
The whole is 1100.	
	_____
b. $255 is 29 percent of what number?	**b.** 950 tons of trash is 59% of what number of tons?
The whole is the unknown part of the problem.	
	_____
c. What percent of 6500 tons is 275 tons?	**c.** What percent of 2650 children attend preschool?
The whole is 6500.	
	_____

Review these examples for Objective 5:	**Now Try:**
5. Identify the parts in the following. Then set up the percent proportion. Do not solve the proportions.	5. Set up the percent proportion. Do not solve the proportions.
a. 72% of 800 children is 576 children.	
First find the percent and the whole.	**a.** Find the amount of sales tax by multiplying $650 and 7.75 percent.
72 is the percent.	
800 is the whole.	
576 is the part.	
Set up the proportion.	
$$\frac{576}{800} = \frac{72}{100}$$	_____

b. $255 is 29% of what number?

First find the percent and the whole.
   29 is the percent.
   whole is unknown.
   255 is the part.
Set up the proportion.
$$\frac{255}{\text{unknown}} = \frac{29}{100}$$

b. 83% of 9000 students is 7470.

_____

c. 95% of 650 is what number?

First find the percent and the whole.
   95 is the percent.
   650 is the whole.
   part is unknown.
Set up the proportion.
$$\frac{\text{unknown}}{650} = \frac{95}{100}$$

c. 343 is 35% of what number.

_____

Objective 1   Learn the percent proportion.

For extra help, see page 404 of your text and Section Lecture video for Section 6.3.

1.   Write the percent proportion

1. _____

Objective 2   Solve for an unknown value in a percent proportion.

For extra help, see Example 1 on pages 404–405 of your text and Section Lecture video for Section 6.3 and Exercise Solutions Clip 9, 11, and 15.

Use the percent proportion to solve for the unknown value. Round to the nearest tenth, if necessary. If the answer is a percent, be sure to include a percent sign.

2.   part = 18, percent = 150

2. _____

3.   whole = 50, percent = 175

3. _____

4.   part = 160, whole = 120

4. _____

Objective 3   Identify the percent.

For extra help, see Example 2 on page 406 of your text and Section Lecture video for Section 6.3 and Exercise Solutions Clip 21.

Identify the percent. Do not try to solve for any unknowns.

5.   83% of what number is 21.5?   5. _____

6.   36 is 72% of what number?   6. _____

7.   17% of Tom's check of $340 is withheld. How much   7. _____
is withheld?

Objective 4   Identify the whole.

For extra help, see Example 3 on page 407 of your text and Section Lecture video for Section 6.3 and Exercise Solutions Clip 31.

Identify the whole. Do not try to solve for any unknowns.

8.   71 is what percent of 384?   8. _____

9.   0.68% of 487 is what number?   9. _____

10.   What is 14% of 78?   10. _____

Objective 5   Identify the part.

For extra help, see Example 4 on page 407 of your text and Section Lecture video for Section 6.3.

Identify the part. Do not try to solve for any unknowns.

11.   29.81 is what percent of 508?   11. _____

12.   16.74 is 11.9% of what number?   12. _____

13.   What number is 12.4% of 1408?   13. _____

Chapter 6 PERCENT

6.4   Using Proportions to Solve Percent Problems

Learning Objectives
1   Use the percent proportion to find the part.
2   Find the whole using the percent proportion.
3   Find the percent using the percent proportion.

Key Terms

Use the vocabulary terms listed below to complete each statement in exercises 1–2.

        cross products   **percent proportion**

1.   Solve a proportion using _____.

2.   The equation $\dfrac{\text{part}}{\text{whole}} = \dfrac{\text{percent}}{100}$ is called the _____.

Guided Examples

Review these examples for Objective 1:

1.   Find 24% of $650.

     Here the percent is 24 and the whole is $650. Now find the part. Let x represent the unknown part.

$$\frac{x}{650} = \frac{24}{100} \quad \text{or} \quad \frac{x}{650} = \frac{6}{25}$$

     Find the cross products in the proportion and show that they are equal.

$$x \cdot 25 = 650 \cdot 6$$

$$x \cdot 25 = 3900$$

$$\frac{x \cdot 25}{25} = \frac{3900}{25}$$

$$x = 156$$

     24% of $650 is $156.

Now Try:

1.   Find 38% of 6500.

                   _____

Name: Date:
Instructor: Section:

2. Use multiplication to find the part.

a. Find 74% of 580 ft.

Step 1 Here the percent is 74. Write 74% as the decimal 0.74.
Step 2 Multiply 0.74 and the whole, which is 580.

$$(0.74)(580) = 429.2$$

It is a good idea to estimate the answer.

$$(0.7)(600) = 420$$

So the exact answer of 429.2 is reasonable.

b. Find 20% of 1980 trucks.

Identify the percent as 20. Write 20% in decimal form as 0.20. Now multiply 0.20 and 1980.

$$(0.20)(1980) = 396$$

You can also use a shortcut to find the answer. Since 20% means 20 parts out of 100 parts, this is the same as $\frac{1}{5}$ of the whole. You can find $\frac{1}{5}$ of a number by dividing the number by 5.
So, this shortcut gives us the exact answer,

$$1980 \div 5 = 396.$$

c. Find 125% of 80 miles.

In this problem, the percent is 125. Write 125% as the decimal 1.25. Next, multiply 1.25 and 80.

$$(1.25)(80) = 100 \text{ miles}$$

d. Find 0.6% of 20 meters.

$$\text{part} = (0.006)(20) = 0.12 \text{ meters}$$

Estimate the answer by realizing that 0.6% is less than 1%.

1% of 20 meters = 0.2 meters

So, our exact answer should be less than 0.2 meters, and 0.12 meters fits this requirement.

3. The freshman class in Central High School has 520 students. Of these students, 15% are in the marching band. How many students are in the marching band? Use the six-problem solving steps.

Step 1 Read the problem. The problem asks us to find the number of students who are in the

2. Use multiplication to find the part.
a. 95% of 3200 pens

_____

b. 62% of 150 miles

_____

c. 175% of 70 gallons

_____

d. 0.8% of $320

_____

3. On Friday afternoon, 450 cars exit a parking garage. If 22% have their lights on when exiting, how many exiting cars have their lights on? Use the six-problem solving steps.

_____

marching band.

Step 2 Work out a plan. Look at the word of as an indicator word for multiplication.
The total number of students is 520, so the whole is 520. The percent is 15. To find the number of students in the marching band, find the part.

Step 3 Estimate a reasonable answer. You can estimate the answer by rounding 15% to 20% and 520 to 500. Remember that 20% is 20 parts out of 100, which is equivalent to $\frac{1}{5}$. So, divide 500 by 5.

$500 \div 5 = 100$ students in marching band

Step 4 Solve the problem.
part $= (0.15)(520) = 78$

Step 5 State the answer. There are 78 students in marching band.

Step 6 Check. The exact answer 78 students in marching band is close to our estimate of 100.

Review these examples for Objective 2:

4.

a. 12 pizzas is 3% of what number of pizzas?

Here the percent is 3, the whole is unknown, and the part is 12. Use the percent proportion to find the whole. Let x represent the unknown whole.

$$\frac{\text{part}}{\text{whole}} = \frac{\text{percent}}{100} \quad \text{so} \quad \frac{12}{x} = \frac{3}{100}$$

$$x \cdot 3 = 12 \cdot 100$$

$$x \cdot 3 = 1200$$

$$\frac{x \cdot \overset{1}{\cancel{3}}}{\underset{1}{\cancel{3}}} = \frac{1200}{3}$$

$$x = 400$$

12 pizzas is 3% of 400 pizzas.

Now Try:

4.

a. 319 cups of coffee is 22% of what number of cups of coffee?

_____

b. 117 books is 18% of what number of books?

The percent is 18 and the part is 117.

$$\frac{117}{x} = \frac{18}{100}$$

$$\frac{117}{x} = \frac{9}{50}$$

$$x \cdot 9 = 117 \cdot 50$$

$$x \cdot 9 = 5850$$

$$\frac{x \cdot \overset{1}{\cancel{9}}}{\underset{1}{\cancel{9}}} = \frac{5850}{9}$$

$$x = 650$$

117 books is 18% of 650 books.

b. 399 miles is 42% of what number of miles?

_____

5. Kathy's overtime pay is $420, which is 12% of her total pay. What is her total pay?

Step 1 Read the problem. The problem asks for the total pay.

Step 2 Work out a plan. From the information in the problem, the percent is 12 and the part from the total pay is $420. The total pay, which is the whole, is the unknown.

Step 3 Estimate a reasonable answer. Round the overtime pay from $420 to $400. Then round 12% to 10%, which is equivalent to $\frac{1}{10}$, and $400 is $\frac{1}{10}$ of the total pay.

$$\$400 \cdot 10 = \$4000$$

Step 4 Solve the problem. Use the percent proportion to find the whole (the total pay).

$$\frac{420}{x} = \frac{12}{100}$$

$$\frac{420}{x} = \frac{3}{25}$$

$$x \cdot 3 = 420 \cdot 25$$

$$x \cdot 3 = 10,500$$

$$\frac{x \cdot \overset{1}{\cancel{3}}}{\underset{1}{\cancel{3}}} = \frac{10,500}{3}$$

$$x = 3500$$

Step 5 State the answer. The total pay is $3500.

Step 6 Check. The exact answer, $3500 is close to the estimate of $4000.

5. In one chemistry class, 60% of the students passed. If 90 students passed, how many students were in the class?

_____

Name: _____   Date: _____
Instructor: _____   Section: _____

Review these examples for Objective 3:

6.

   a. What percent of 8000 is 4?

   The whole is 8000 and the part is 4. Next, find the percent.

$$\frac{\text{part}}{\text{whole}} = \frac{\text{percent}}{100} \quad \text{so} \quad \frac{4}{8000} = \frac{x}{100}$$

$$\frac{1}{2000} = \frac{x}{100}$$

   Find the cross products.

$$2000 \cdot x = 1 \cdot 100$$

$$2000 \cdot x = 100$$

$$\frac{\overset{1}{\cancel{2000}} \cdot x}{\underset{1}{\cancel{2000}}} = \frac{100}{2000}$$

$$x = 0.05$$

   4 is 0.05% of 8000.

   b. 550 is what percent of 1000?

   The whole is 1000 and the part is 550.

$$\frac{550}{1000} = \frac{x}{100}$$

$$\frac{11}{20} = \frac{x}{100}$$

$$20 \cdot x = 11 \cdot 100$$

$$20 \cdot x = 1100$$

$$\frac{\overset{1}{\cancel{20}} \cdot x}{\underset{1}{\cancel{20}}} = \frac{1100}{20}$$

$$x = 55$$

   550 is 55% of 1000.

7. G&G Pharmacy has a total payroll of $89,350, of which $19,657 goes towards employee fringe benefits. What percent of the total payroll goes to fringe benefits?

Step 1 Read the problem. The problem asks for the percent of the total payroll that goes to employee fringe benefits.

Step 2 Work out a plan. The total payroll is the whole, which is $89,350. The part is $19,657. Use the percent proportion to find the percent of

Now Try:

6.

   a. What percent of 16 is 2?

   _____

   b. 650 is what percent of 13?

   _____

7. In one shipment, 6950 out of 257,800 crates were damaged. What percent of the crates were damaged?

   _____

Copyright © 2014 Pearson Education, Inc.

the total payroll that goes to fringe benefits.

Step 3 Estimate a reasonable answer. Round the total payroll from $89,350 to $90,000. Then round $19,657 to $20,000. Divide to find the estimate.

$$\frac{20,000}{90,000} \approx 22.2\%$$

Step 4 Solve the problem. Let x represent the unknown percent.

$$\frac{19,657}{89,350} = \frac{x}{100}$$
$$89,350 \cdot x = 19,657 \cdot 100$$
$$89,350 \cdot x = 1,965,700$$
$$\frac{\overset{1}{\cancel{89,350}} \cdot x}{\underset{1}{\cancel{89,350}}} = \frac{1,965,700}{89,350}$$
$$x = 22$$

Step 5 State the answer. 22% of the total payroll goes towards fringe benefits.

Step 6 Check. The exact answer of 22% is close to the estimate of 22.2%.

8. The budget for a political fund was $2.5 million. This year $3.5 million was given to the fund. What percent of the budget was given to the fund?

Step 1 Read the problem. The problem asks for the percent of the budget given to the fund.

Step 2 Work out a plan. The budget is the whole, which is $2.5 million. The contribution this year is the budget and more, or $3.5 million (part = 3.5). You need to find the percent of the budget that was given to the fund.

Step 3 Estimate a reasonable answer. The increase is $1 million and the whole is $3 million. The increase is $\frac{1}{3}$ which is about 33%. The whole is 100%, so 100% + 33% = 133%, our estimate.

Step 4 Solve the problem. Let x represent the unknown percent.

8. A marathon committee planned for 4200 participants. There were 5082 runners who registered. What percent of the planned participants registered?

_____

Name: Date:
Instructor: Section:

$$\frac{3.5}{2.5} = \frac{x}{100}$$

$$\frac{7}{5} = \frac{x}{100}$$

$$5 \cdot x = 7 \cdot 100$$

$$5 \cdot x = 700$$

$$\frac{\overset{1}{\cancel{5}} \cdot x}{\underset{1}{\cancel{5}}} = \frac{700}{5}$$

$$x = 140$$

Step 5 State the answer. The amount given to the fund is 140% of the budget.

Step 6 Check. The exact answer, 140% is close to the estimate of 133%.

Objective 1 Use the percent proportion to find the part.

For extra help, see Examples 1–3 on pages 412–414 of your text and Section Lecture video for Section 6.4 and Exercise Solutions Clip 5, 13, and 43a..

Use the percent proportion to find the part. Round to the nearest tenth, if necessary.

1. 20% of 1400

1. _____

Use multiplication to find the part. Round to the nearest tenth, if necessary.

2. 175% of 50

2. _____

Solve the application problem. Round to the nearest tenth, if necessary.

3. A survey at an intersection found that of 2200 drivers, 43% were wearing seat belts. How many drivers in the survey were wearing seat belts?

3. _____

Objective 2   Find the whole using the percent proportion.

For extra help, see Examples 4–5 on pages 414–415 of your text and Section Lecture video for Section 6.4 and Exercise Solutions Clip 21.

Use the percent proportion to find the whole. Round to the nearest tenth, if necessary.

4.   36% of what number is 75?   **4.** _____

5.   550 is 110% of what number?   **5.** _____

Solve the application problem. Round to the nearest tenth, if necessary.

6.   This year, there are 960 scholarship applications,   **6.** _____
which is 120% of the number of applications last
year. Find the number of applications last year.

Objective 3   Find the percent using the percent proportion.

For extra help, see Examples 6–8 on pages 416–418 of your text and Section Lecture video for Section 6.4.

Use the percent proportion to find the whole. Round to the nearest tenth, if necessary.

7.   7 is what percent of 280?   **7.** _____

8.   What percent of 4.5 is 3.9?   **8.** _____

Solve the application problem. Round to the nearest tenth, if necessary.

9.   In a motor cross, the leader has completed 108.8   **9.** _____
miles of the 128-mile course. What percent of the
total course has she completed?

Chapter 6 PERCENT

6.5   Using the Percent Equation

Learning Objectives
1   Use the percent equation to find the part.
2   Find the whole using the percent equation.
3   Find the percent using the percent equation.

Key Terms

Use the vocabulary terms listed below to complete each statement in exercises 1–2.

       percent equation   **percent**

1.   A number written with a _____ sign means "divided by 100".

2.   The _____ is part $=$ percent $\times$ whole.

Guided Examples

Review these examples for Objective 1:	**Now Try:**
1.	**1.**
a. Find 16% of $2300.	**a.** Find 18% of $350.

Write 16% as the decimal 0.16. The whole, which comes after the word of, is 2300. Next, use the percent equation. Let x represent the unknown part.
 	
    part $=$ percent $\times$ whole
 	
      $x = (0.16)(2300)$
 	
Multiply 0.16 and 2300.
    $x = 368$
16% of $2300 is $368.
	_____

b. Find 120% of 90 cartons.	**b.** Find 130% of 70 packages.

Write 120% as the decimal 1.20. The whole is 90. Let x represent the unknown part.
 	
    part $=$ percent $\times$ whole
 	
      $x = (1.20)(90)$
 	
      $x = 108$
120% of 90 cartons is 108 cartons.
	_____

c. Find 0.8% of 3500 cases.

Write 0.8% as the decimal 0.008. The whole is 3500. Let x represent the unknown part.

$$\text{part} = \text{percent} \times \text{whole}$$

$$x = (0.008)(3500)$$

$$x = 28$$

0.8% of 3500 cases is 28 cases.

c. Find 0.6% of 31,000 students.

_____

Review these examples for Objective 2:	**Now Try:**
2.	**2.**

a. 160 gallons is 25% of what number of gallons?

The part is 160 and the percent is 25% or the decimal 0.25. The whole is unknown.

160 is 25% of what number?

Now, use the percent equation.

$$\text{part} = \text{percent} \times \text{whole}$$

$$160 = (0.25)(x)$$

$$\frac{160}{0.25} = \frac{\overset{1}{\cancel{(0.25)}}(x)}{\underset{1}{\cancel{0.25}}}$$

$$640 = x$$

160 gallons is 25% of 640 gallons.

a. 45 units is 37.5% of what number of units.

_____

b. 25 employees is 5% of what number of employees.

Write 5% as 0.05. The part is 25. Use the percent equation to find the whole.

$$\text{part} = \text{percent} \times \text{whole}$$

$$25 = (0.05)(x)$$

$$\frac{25}{0.05} = \frac{\overset{1}{\cancel{(0.05)}}(x)}{\underset{1}{\cancel{0.05}}}$$

$$500 = x$$

25 employees is 5% of 500 employees.

b. 330 points is 13.25% of how many points

_____

c. 75 is $6\frac{1}{4}\%$ of what number?

Write $6\frac{1}{4}\%$ as 6.25%, or the decimal 0.0625.

The part is 75. Use the percent equation.

c. 1750% of what number is 1050?

_____

$$\text{part} = \text{percent} \times \text{whole}$$

$$75 = (0.0625)(x)$$

$$\frac{75}{0.0625} = \frac{\overset{1}{\cancel{(0.0625)}}(x)}{\underset{1}{\cancel{0.0625}}}$$

$$1200 = x$$

75 is $6\frac{1}{4}\%$ of 1200.

Review these examples for Objective 3:	**Now Try:**
3.	**3.**

a. $700 is what percent of $2800?

Because $2800 follows of, the whole is $2800. The part is $700, and the percent is unknown. Use the percent formula.

$$\text{part} = \text{percent} \cdot \text{whole}$$

$$700 = x \cdot 2800$$

$$\frac{700}{2800} = \frac{x \cdot \overset{1}{\cancel{2800}}}{\underset{1}{\cancel{2800}}}$$

$$0.25 = x$$

$700 is 25% of $2800.

a. 145 miles is what percent of 500 miles?

_____

b. What percent of 250 is 115?

The whole is 250 and the part is 115. Let x represent the unknown percent.

$$\text{part} = \text{percent} \cdot \text{whole}$$

$$115 = x \cdot 250$$

$$\frac{115}{250} = \frac{x \cdot \overset{1}{\cancel{250}}}{\underset{1}{\cancel{250}}}$$

$$0.46 = x$$

46% of 250 is 115.

b. What percent of $225,000 is $45,000?

_____

c. What percent of 1500 calories is 1800 calories?

The whole is 1500 and the part is 1800. Let x represent the unknown percent.

$$\text{part} = \text{percent} \cdot \text{whole}$$

$$1800 = x \cdot 1500$$

$$\frac{1800}{1500} = \frac{x \cdot \cancel{1500}^{1}}{\cancel{1500}_{1}}$$

$$1.2 = x$$

1.2 is 120%

120% of 1500 calories is 1800 calories.

d. $2400 is what percent of $35,000?

Since $35,000 follows of, the whole is $35,000. The part is $2400.

$$\text{part} = \text{percent} \cdot \text{whole}$$

$$2400 = x \cdot 35,000$$

$$\frac{2400}{35,000} = \frac{x \cdot \cancel{35,000}^{1}}{\cancel{35,000}_{1}}$$

$$0.07 = x$$

7% of $35,000 is $2450.

c. What percent of 90 is 288?

_____

d. 24 miles is what percent of 6000 miles?

_____

Objective 1   Use the percent equation to find the part.

For extra help, see Example 1 on pages 425–426 of your text and Section Lecture video for Section 6.5 and Exercise Solutions Clip 5, 9, and 13.

Find the part using the percent equation. Round to the nearest tenth, if necessary.

1.   9% of 240

2.   140% of 76

3.   0.4% of 350

1. _____

2. _____

3. _____

Objective 2   Find the whole using the percent equation.

For extra help, see Example 2 on pages 426–427 of your text and Section Lecture video for Section 6.5 and Exercise Solutions Clip 21.

Find the whole using the percent equation. Round to the nearest tenth, if necessary.

4.   64 is 40% of what number?   4. _____

5.   75% of what number is 1125?   5. _____

6.   35 is 153% of what number?   6. _____

Objective 3   Find the percent using the percent equation.

For extra help, see Example 3 on pages 427–428 of your text and Section Lecture video for Section 6.5 and Exercise Solutions Clip 27 and 31.

Find the percent using the percent equation. Round to the nearest tenth, if necessary.

7.   15 is what percent of 75?   7. _____

8.   What percent of 160 is 8?   8. _____

9.   What percent of 18 is 44?   9. _____

Chapter 6 PERCENT

6.6   Solving Application Problems with Percent

Learning Objectives
1   Find sales tax.
2   Find commissions.
3   Find the discount and sale price.
4   Find the percent of change.

Key Terms

Use the vocabulary terms listed below to complete each statement in exercises 1–4.

sales tax   **commission**   **discount**

percent of increase or decrease

1.   _____ is a percent of the dollar value of total sales paid to a
      salesperson.

2.   In a _____ problem, the increase or decrease is a
      percent of the original amount.

3.   The percent of the total sales charged as tax is called the _____.

4.   The percent of the original price that is deducted from the original price is called
      the _____.

Guided Examples

Review these examples for Objective 1:

1.   A refrigerator sells for $595. If the sales tax rate
      is 6%, how much tax is paid? What is the total
      cost?

      Step 1 Read the problem. The problem asks for
      the total cost of the refrigerator, including sales
      tax.

      Step 2 Work out a plan. Use the sales tax
      formula to find the amount of sales tax. Write
      the tax rate (6%) as a decimal (0.06). The cost of
      the item is $595. Use the letter *a* to represent the
      unknown amount of tax. Add the sales tax to the
      cost of the item.

      Step 3 Estimate a reasonable answer. Round
      $595 to $600. Round 6% to 5%. Recall that 5%

Now Try:

1.   A television set sells for $750
      plus 8% sales tax. Find the price
      of the TV including sales tax.

      _____

is equivalent to $\dfrac{1}{20}$, so divide $600 by 20 to estimate the tax.

$600 \div 20 = 30 tax

The total estimated cost is $600 + $30 = $630.

Step 4 Solve the problem.

$$\underbrace{\text{part}}_{\text{amount of sales tax}} = \underbrace{\text{percent}}_{\text{rate of tax}} \cdot \underbrace{\text{whole}}_{\text{cost of item}}$$

$$a = (6\%)(\$595)$$
$$a = (0.06)(\$595)$$
$$a = \$35.70$$

The tax paid on the refrigerator is $35.70. The customer would pay a total cost of $595 + $35.70 = $630.70.

Step 5 State the answer. The total cost of the refrigerator is $630.70.

Step 6 Check. The exact answer $630.70 is close to our estimate of $630.

2. The sales tax for a $320 office chair is $25.60. Find the rate of sales tax.

    Step 1 Read the problem. The problem asks for the sales tax rate.

    Step 2 Work out a plan. Use the sales tax formula.

    sales tax = rate of tax · cost of item

    Solve for the rate of tax, which is the percent. The cost of the office chair (the whole) is $320, and the amount of sales tax (the part) is $25.60. Use r to represent the unknown rate of tax (the percent).

    Step 3 Estimate a reasonable answer. Round $320 to $300, and round $25.60 to $30. The sales tax is $\dfrac{30}{300}$ or $\dfrac{1}{10}$ of the cost of the office chair.

    $$\dfrac{1}{10} = 10\%$$

    Step 4 Solve the problem.

    sales tax = rate of tax · cost of item

2. The sales tax for a $78 utility bill is $1.17. Find the rate of the sales tax.

    _____

$$\$25.60 = r \cdot \$320$$

$$\frac{25.60}{320} = \frac{r \cdot \cancel{320}^{1}}{\cancel{320}_{1}}$$

$$0.08 = r$$

0.08 is 8%

Step 5 State the answer. The sales tax rate is 8%.

Step 6 Check. The exact answer, 8%, is close to our estimate of 10%.

Review these examples for Objective 2:	**Now Try:**
3. Nicole had sales of $18,306 in the month of October. If her rate of commission is 12%, find the amount of commission that she earned.	3. Bill had sales of $156,000 last year. If his commission rate is 3%, find the amount of his commission.

Step 1 Read the problem. The problem asks for the amount of commission that Nicole earned.

Step 2 Work out a plan. Use the commission formula. Write the rate of commission (12%) as a decimal (0.12). The amount of Nicole's sales ($18,306) is the whole. Use c to represent the unknown amount of commission.

Step 3 Estimate a reasonable answer. Round the commission rate of 12% to 10%. Round the amount of sales from $18,306 to $20,000. Since 10% is equivalent to $\frac{1}{10}$, divide $20,000 by 10 to estimate the amount of commission.

$$\$20,000 \div 10 = \$2000$$

Step 4 Solve the problem.

amount of commission = rate of commission · amount of sales

$$c = (12\%)(\$18,306)$$

$$c = (0.12)(\$18,306)$$

$$c = \$2196.72$$

Step 5 State the answer. Nicole earned a commission of $2196.72.

Step 6 Check. The exact answer, $2196.72, is close to our estimate of $2000.

4. A business property has just been sold for $1,692,804. The real estate agent selling the property earned a commission of $42,320. Find the rate of commission.

Step 1 Read the problem. In this problem, we must find the rate (percent) of commission.

Step 2 Work out a plan. You could use the commission formula. Another approach is to use the percent proportion. The whole is $1,692,804, the part is $42,320, and the percent is unknown.

Step 3 Estimate a reasonable answer. Round the commission, $42,320 to $40,000, and round $1,692,804 to $2,000,000. The commission in fraction form is $\dfrac{\$40,000}{\$2,000,000}$, which simplifies to $\dfrac{1}{50}$. Changing $\dfrac{1}{50}$ to a percent gives 2%, as our estimate.

Step 4 Solve the problem.

$$\frac{\text{part}}{\text{whole}} = \frac{x}{100}$$

$$\frac{42,320}{1,692,804} = \frac{x}{100}$$

$$1,692,804 \cdot x = 42,320 \cdot 100$$

$$\frac{\overset{1}{\cancel{1,692,804}} \cdot x}{\underset{1}{\cancel{1,692,804}}} = \frac{4,232,000}{1,692,804}$$

$$x = 2.5$$

Step 5 State the answer. The rate of commission is 2.5%.

Step 6 Check. The exact answer of 2.5% is close to our estimate of 2%.

4. Susan earned a commission of $3750 for selling $25,000 worth of materials. Find the rate of commission.

_____

Review this example for Objective 3:

5. Mike Lee can purchase a new car at 8% below window sticker price. Find the sale price on a car with a window sticker price of $17,600.

Step 1 Read the problem. The problem asks for the price of the car after a discount of 8%.

Step 2 Work out a plan. The problem is solved in two steps. First, find the amount of the

Now Try:

5. A hard-cover book with an original price of $24.95 is on sale at 60% off. Find the sale price of the book.

_____

discount, that is, the amount that will be "taken off" (subtracted), by multiplying the original price ($17,600) by the rate of the discount (8%). The second step is to subtract the amount of discount from the original price. This gives the sale price, which is what Mike will actually pay for the car.

Step 3 Estimate a reasonable answer. Round the original price from $17,600 to $20,000, and the rate of discount from 8% to 10%. Since 10% is equivalent to $\frac{1}{10}$, the estimated discount is $20,000 \div 10 = \$2000$, so the estimated sale price is $20,000 - \$2000 = \$18,000$.

Step 4 Solve the problem. First find the exact amount of the discount.

amount of discount = rate of discount · original price

$$a = (8\%)(\$17,600)$$
$$a = (0.08)(\$17,600)$$
$$a = \$1408$$

Now, find the sale price of the car by subtracting the amount of the discount ($1408) from the original price.

sale price = original price − amount of discount
$$= \$17,600 - \$1408$$
$$= \$16,192$$

Step 5 State the answer. The sale price of the car is $16,192.

Step 6 Check. The exact answer, $16,192, is close to the estimate of $18,000.

Review these examples for Objective 4:

6. The membership of Pleasant Acres Golf Club was 320 two years ago. The membership is now 740. Find the percent of increase in the two years.

Step 1 Read the problem. The problem asks for the percent of increase.

Step 2 Work out a plan. Subtract the membership two years ago (320) from the attendance now (740) to find the amount of increase in membership. Next, use the percent proportion. The whole is 740 (the membership

Now Try:

6. Enrollment in secondary education courses increased from 1900 students last semester to 2280 students this semester. Find the percent of increase.

_____

now), the part is 420 (the amount of increase in membership), and the percent is unknown.

Step 3 Estimate a reasonable answer. Round 320 to 300, and 740 to 700. The amount of increase is 700 − 300 = 400. Since 400 (the increase) is one and a third, or 133% as large as the original amount, the estimated percent of increase is 133%.

Step 4 Solve the problem.
    740 − 320 = 420

$$\frac{420}{320} = \frac{x}{100}$$

Solve this proportion to find that $x = 131.25$.

Step 5 State the answer. The percent increase is 131.25%.

Step 6 Check. The exact answer, 131.25%, is close to our estimate of 133%.

7. The number of days employees of Prodex Manufacturing Company were absent from their jobs decreased from 96 days last month to 72 days this month. Find the percent of decrease.

Step 1 Read the problem. The problem asks for the percent of decrease.

Step 2 Work out a plan. Subtract the number of days from this month (72) from the number of days last month (96) to find the amount of decrease. Then, use the percent proportion. The whole is 96 (last month's original number of days), the part is 24 (amount of decrease of days), and the percent is unknown.

Step 3 Estimate a reasonable answer. Estimate the answer by rounding 96 to 100, and 72 to 70.

The decrease is 100 − 70 = 30. Since 30 is $\frac{3}{10}$ of 100, our estimate is $30 \div 100 = 0.3$ or 30%.

Step 4 Solve the problem.
    96 − 72 = 24

$$\frac{24}{96} = \frac{x}{100}$$

Solve this proportion to find that $x = 25$.

7. One day in 2008, the Dow Jones Industrial Average dropped from about 12,635 to 12,265. Find the percent of decrease.

_____

Step 5 State the answer. The percent decrease is 25%.

Step 6 Check. The exact answer, 25% is close to our estimate of 30%.

Objective 1   Find sales tax.

For extra help, see Examples 1–2 on pages 435–436 of your text and Section Lecture video for Section 6.6.

Find the amount of sales tax and the total cost. Round answers to the nearest cent, if necessary.

	Amount of sale	Tax Rate
1.	$50	7%

1. Tax _____

   Total _____

Find the sales tax rate. Round answers to the hundredth, if necessary.

	Amount of sale	Amount of Tax
2.	$450	$36

2. _____

Solve the application problem. Round money answers to the nearest cent, if necessary.

3. A gold bracelet costs $1300 not including a sales tax of $71.50. Find the sales tax rate.

3. _____

Objective 2   Find commissions.

For extra help, see Examples 3–4 on pages 437–438 of your text and Section Lecture video for Section 6.6 and Exercise Solutions Clip 11.

Find the commission earned. Round answers to the nearest cent, if necessary.

	Amount of sale	Rate of Commission
4.	$75,000	4%

4. _____

Find the rate of commission. Round answers to the hundredth, if necessary.

	Amount of Sale	Amount of Commission
5.	$3200	$480

5. _____

Objective 3   Find the discount and sale price.

For extra help, see Example 5 on pages 438 of your text and Section Lecture video for Section 6.6 and Exercise Solutions Clip 21 and 23.

Find the amount of discount and the amount paid after the discount. Round money answers to the nearest cent, if necessary.

	Original price	**Rate of Discount**	
6.	$200	15%	**6. Discount** _____
			Amount paid _____

Solve each application problem. Round money answers to the nearest cent, if necessary.

7.   A "Super 35% Off Sale" begins today. What is the price of a hair dryer normally priced at $15?

7. _____

8.   Geishe's Shoes sells shoes at 33% off the regular price. Find the price of a pair of shoes normally priced at $54, after the discount is given.

8. _____

Objective 4   Find the percent of change.

For extra help, see Examples 6–7 on pages 439–440 of your text and Section Lecture video for Section 6.6.

Solve each application problem. Round to the nearest tenth of a percent, if necessary.

9.   The earnings per share of Amy's Cosmetic Company decreased from $1.20 to $0.86 in the last year. Find the percent of decrease.

9. _____

10.   The price of a certain model of calculator was $33.50 five years ago. This calculator now costs $18.75. Find the percent of decrease in the price in the last five years.

10. _____

11.   In 1980, there were approximately 3,612,000 births in the U.S. In 2002, there were approximately 4,022,000 births in the U.S. Find the percent of increase.

11. _____

Chapter 6 PERCENT

6.7   Simple Interest

Learning Objectives

1   Find the simple interest on a loan.
2   Find the total amount due on a loan.

Key Terms

Use the vocabulary terms listed below to complete each statement in exercises 1–5.

      interest   **interest formula**   **simple interest**   **principal**

      rate of interest

1. The charge for money borrowed or loaned, expressed as a percent, is called
   _____.

2. A fee paid for borrowing or lending money is called _____.

3. The formula $I = p \cdot r \cdot t$ is called the _____.

4. Use the formula $I = p \cdot r \cdot t$ to compute the amount of _____
   due on a loan.

5. The amount of money borrowed or loaned is called the _____.

Guided Examples

Review these examples for Objective 1:

1. Find the interest on $4000 at 4% for 1 year.

   The amount borrowed, or principal (p), is $4000. The interest rate (r), is 4%, which is 0.04 as a decimal, and the time of the loan (t) is 1 year. Use the formula.

   $I = p \cdot r \cdot t$

   $I = (4000)(0.04)(1)$

   $I = \$160$

   The interest is $160.

2. Find the interest on $5600 at 5% for two and a half years.

   The principal (p) is $5600. The rate ($r$) is 5%, or 0.05 as a decimal, and the time (t) is $2\frac{1}{2}$ or 2.5 years. Use the formula.

Now Try:

1. Find the interest on $80 at 5% for 1 year.

   _____

2. Find the interest on $620 at 16% for one and a quarter years.

   _____

$$I = p \cdot r \cdot t$$
$$I = (5600)(0.05)(2.5)$$
$$I = \$700$$

The interest is $700.

3. Find the interest on $720 at $3\frac{1}{2}\%$ for 8 months.

The principal is $720. The rate is $3\frac{1}{2}\%$ or 0.035

as a decimal, and the time is $\frac{8}{12}$ of a year. Use

the formula $I = p \cdot r \cdot t$.

$$I = (720)(0.035)\left(\frac{8}{12}\right)$$
$$= 25.2\left(\frac{2}{3}\right)$$
$$= 16.80$$

The interest is $16.80.

3. Find the interest on $840 at $8\frac{1}{2}\%$ for 5 months.

_____

Review this example for Objective 2:

4. A loan of $1500 was made at 12% interest for 10 months. Find the total amount due.

First find the interest. Then add the principal and the interest to find the total amount due.

$$I = (1500)(0.12)\left(\frac{10}{12}\right)$$
$$= \$150$$

The interest is $150.

     Total amount due = principal + interest
$$= \$1500 + \$150$$
$$= \$1650$$

The total amount due is $1650.

Now Try:

4. Mary Ann borrows $1200 at 10% for 4 months. Find the total amount due.

_____

Name: Date:
Instructor: Section:

Objective 1   Find the simple interest on a loan.

For extra help, see Examples 1–3 on pages 445–446 of your text and Section Lecture video for Section 6.7 and Exercise Solutions Clip 7 and 21.

Find the interest. Round to the nearest cent, if necessary.

	Principal	**Rate**	**Time in Years**	
1.	$5280	9%	1	1. _____
2.	$780	10%	$2\frac{1}{2}$	2. _____
3.	$14,400	7%	7 months	3. _____

Objective 2   Find the total amount due on a loan.

For extra help, see Example 4 on page 446 of your text and Section Lecture video for Section 6.7.

Find the total amount due on each loan. Round to the nearest cent, if necessary.

	Principal	**Rate**	**Time**	
4.	$200	11%	1 year	4. _____
5.	$3000	5%	6 months	5. _____
6.	$900	10%	$2\frac{1}{2}$ years	6. _____

Chapter 6 PERCENT

6.8   Compound Interest

Learning Objectives
1   Understand compound interest.
2   Understand compound amount.
3   Find the compound amount.
4   Use a compound interest table.
5   Find the compound amount and the amount of interest.

Key Terms

Use the vocabulary terms listed below to complete each statement in exercises 1–3.

   compound interest   **compound amount**   **compounding**

1.   Interest paid on principal plus past interest is called _____.

2.   The total amount in an account, including compound interest and the original principal, is called the _____.

3.   When the amount of interest is computed based on the principal plus the past interest, use a process called _____.

Guided Examples

Review this example for Objective 2:
1.   Skylar deposits $4600 in an account that pays 2% interest compounded annually for 4 years. Find the compounded amount. Round to the nearest cent when necessary.

Year 1:
Interest: $(\$4600)(0.02)(1) = \$92$
Compound amount: $\$4600 + \$92 = \$4692$

Year 2:
Interest: $(\$4692)(0.02)(1) = \$93.84$
Compound amount: $\$4692 + \$93.84 = \$4785.84$

Year 3:
Interest: $(\$4785.84)(0.02)(1) \approx \$95.72$
Compound amount:
    $\$4785.84 + \$95.72 = \$4881.56$

Year 4:
Interest: $(\$4881.56)(0.02)(1) \approx \$97.63$
Compound amount:
    $\$4881.56 + \$97.63 = \$4979.19$

Now Try:
1.   Barbara deposits $4500 in an account that pays 4% interest compounded annually for 2 years. Find the compounded amount. Round to the nearest cent when necessary.

_____

Review this example for Objective 3:

2. Find the compound amount in Example 1 using multiplication.

$$(\$4600)(1.02)(1.02)(1.02)(1.02) \approx \$4979.19$$

Our answer, $4979.19 is the same as in Example 1.

Now Try:

2. Find the compound amount for an account that has $3200 that pays 6% interest compounded annually for 2 years using multiplication.

_____

Time Periods	3.00%	3.50%	4.00%	4.50%	5.00%	5.50%	6.00%	8.00%	Time Periods
1	1.0300	1.0350	1.0400	1.0450	1.0500	1.0550	1.0600	1.0800	1
2	1.0609	1.0712	1.0816	1.0920	1.1025	1.1130	1.1236	1.1664	2
3	1.0927	1.1087	1.1249	1.1412	1.1576	1.1742	1.1910	1.2597	3
4	1.1255	1.1475	1.1699	1.1925	1.2155	1.2388	1.2625	1.3605	4
5	1.1593	1.1877	1.2167	1.2462	1.2763	1.3070	1.3382	1.4693	5
6	1.1941	1.2293	1.2653	1.3023	1.3401	1.3788	1.4185	1.5869	6
7	1.2299	1.2723	1.3159	1.3609	1.4071	1.4547	1.5036	1.7138	7
8	1.2668	1.3168	1.3686	1.4221	1.4775	1.5347	1.5938	1.8509	8
9	1.3048	1.3629	1.4233	1.4861	1.5513	1.6191	1.6895	1.9990	9
10	1.3439	1.4106	1.4802	1.5530	1.6289	1.7081	1.7908	2.1589	10
11	1.3842	1.4600	1.5395	1.6229	1.7103	1.8021	1.8983	2.3316	11
12	1.4258	1.5111	1.6010	1.6959	1.7959	1.9012	2.0122	2.5182	12

Review these examples for Objective 4:

3. Find each compound amount using the compound interest table. Round answers to the nearest cent.

a. $1 is deposited at a 6% interest rate for 8 years

Look down the column headed 6%, and across to row 8 (because 8 years = 8 time periods). At the intersection of the column and row, read the compound amount, 1.5938, which can be rounded to $1.59.

Now Try:

3. Find each compound amount using the compound interest table. Round answers to the nearest cent.

a. $1 is deposited at a 3% interest rate for 7 years.

_____

b. $1 is deposited at a $4\frac{1}{2}\%$ interest rate for 11 years

The intersection of the $4\frac{1}{2}\%$ (4.50%) column and row 11 shows 1.6229 as the compound amount. Round this to $1.62.

b. $1 is deposited at a $5\frac{1}{2}\%$ interest rate for 9 years.

_____

Review these examples for Objective 5:

4. Use the compound interest table to find the compound amount and the interest.

  a. $6000 at 5% for 9 yr

Look in the table for 5% and 9 periods to find the number 1.5513 but do not round it. Multiply this number and the principal of $6000.
   ($6000)(1.5513) = $9307.80
The account will contain $9307.80 after 9 years. Find the interest by subtracting the original deposit from the compound amount.
   $9307.80 – $6000 = $3307.80
A total of $3307.80 in interest was earned.

  b. $9600 at 6% for 5 yrs

Look in the table for 6% and 5 periods to find 1.3382.
   ($9600)(1.3382) = $12,846.72
The account will contain $12,846.72 after 5 years.
Find the interest by subtracting the original deposit from the compound amount.
   $12,846.72 – $9600 = $3246.72
A total of $3246.72 in interest was earned.

Now Try:

4. Use the compound interest table to find the compound amount and the interest.

  a. $1600 at 4% for 6 yr

   _____

  b. $23,000 at 5% for 3 yr

   _____

Objective 1   Understand compound interest.
Objective 2   Understand compound amount.

For extra help, see Example 1 on page 452 of your text and Section Lecture video for Section 6.8 and Exercise Solutions Clip 3 and 5.

   1.   Belinda deposited $2000 in an account earning 5% annually. How much is in the account at the end of the first year?

   1. _____

2. If Belinda makes no withdrawals, how much money   2. _____
is in her account at the end of two years?

3. If Belinda makes no withdrawals, how much money   3. _____
is in her account at the end of three years? How
much interest has she earned in total?   _____

Objective 3   Find the compound amount.

For extra help, see Example 2 on page 452 of your text and Section Lecture video for
Section 6.8 and Exercise Solutions Clip 11 and 13.

Find the compound amount given the following deposits. Interest is compounded
annually. Round to the nearest cent, if necessary.

4. $7000 at 5% for 3 years   4. _____

5. $3200 at 7% for 2 years   5. _____

6. $24,600 at 5% for 4 years   6. _____

Name: Date:
Instructor: Section:

Objective 4   Use a compound interest table.

For extra help, see Example 3 on page 453 of your text and Section Lecture video for Section 6.8 and Exercise Solutions Clip 25.

Use the table for compound interest to find the compound amount. Interest is compounded annually. Round to the nearest cent, if necessary.

Time Periods	3.00%	3.50%	4.00%	4.50%	5.00%	5.50%	6.00%	8.00%	Time Periods
1	1.0300	1.0350	1.0400	1.0450	1.0500	1.0550	1.0600	1.0800	1
2	1.0609	1.0712	1.0816	1.0920	1.1025	1.1130	1.1236	1.1664	2
3	1.0927	1.1087	1.1249	1.1412	1.1576	1.1742	1.1910	1.2597	3
4	1.1255	1.1475	1.1699	1.1925	1.2155	1.2388	1.2625	1.3605	4
5	1.1593	1.1877	1.2167	1.2462	1.2763	1.3070	1.3382	1.4693	5
6	1.1941	1.2293	1.2653	1.3023	1.3401	1.3788	1.4185	1.5869	6
7	1.2299	1.2723	1.3159	1.3609	1.4071	1.4547	1.5036	1.7138	7
8	1.2668	1.3168	1.3686	1.4221	1.4775	1.5347	1.5938	1.8509	8
9	1.3048	1.3629	1.4233	1.4861	1.5513	1.6191	1.6895	1.9990	9
10	1.3439	1.4106	1.4802	1.5530	1.6289	1.7081	1.7908	2.1589	10
11	1.3842	1.4600	1.5395	1.6229	1.7103	1.8021	1.8983	2.3316	11
12	1.4258	1.5111	1.6010	1.6959	1.7959	1.9012	2.0122	2.5182	12

7. $1 at 6% for 7 years   7. _____

8. $1 at 8% for 3 years   8. _____

9. $1 at $5\frac{1}{2}$% for 12 years   9. _____

Name: Date:
Instructor: Section:

Objective 5   Find the compound amount and the amount of interest.

For extra help, see Example 4 on page 454 of your text and Section Lecture video for Section 6.8 and Exercise Solutions Clip 25.

Find the compound amount and the compound interest. Round to the nearest cent, if necessary. Use the table for compound interest in your text book to find the compound amount. Interest is compounded annually.

	Principal	**Rate**	**Time in Years**
10.	$9150	8%	8

10.

Amount _____

Interest_____

| 11. | $21,400 | $4\frac{1}{2}$ % | 11 |

11.
Amount _____

Interest_____

| 12. | $8000 | 3% | 12 |

12.
Amount _____

Interest_____

Chapter 7 MEASUREMENT

7.1   Problem Solving with U.S. Measurement Units

Learning Objectives
1   Learn the basic U.S. measurement units.
2   Convert among U.S. measurement units using multiplication or division.
3   Convert among measurement units using unit fractions.
4   Solve application problems using U.S. measurement units.

Key Terms

Use the vocabulary terms listed below to complete each statement in exercises 1–3.

U.S. measurement units   **unit fractions**   **metric system**

1.   The _____ is based on units of ten.

2.   _____ are used to convert among different measurements.

3.   _____ include inches, feet, quarts, and pounds.

Guided Examples

Review these examples for Objective 1:	**Now Try:**
1. Memorize the U.S. measurement conversions from the text. Then fill in the blanks.	1. Memorize the U.S. measurement conversions from the text. Then fill in the blanks.
a. $2\,c =$ _____ pt	**a.** $16\,oz =$ _____ lb
Answer: 1 pt	
	_____
b. $1\,mi =$ _____ ft	**b.** $1\,gal =$ _____ qt
Answer: 5280 ft	
	_____

Review these examples for Objective 2:	**Now Try:**
2. Convert each measurement.	2. Convert each measurement.
a. 16 yd to feet	**a.** $17\frac{1}{2}$ ft to inches
You are converting from a larger unit to a smaller unit, so multiply. Because 1 yd = 3 ft, multiply by 3. $16\,yd = 16 \cdot 3 = 48$ ft	_____

b. 6 T to pounds

You are converting from a larger unit to a smaller unit, so multiply.
Because 1 T = 2000 lb, multiply by 2000.
$$6\text{ T} = 6 \cdot 2000 = 12{,}000\text{ lb}$$

c. 8 pt to quarts

You are converting from a smaller unit to a larger unit, so divide.
Because 2 pt = 1 qt, divide by 2.
$$8\text{ pt} = \frac{8}{2} = 4\text{ qt}$$

d. 3960 ft to miles

You are converting from a smaller unit to a larger unit, so divide.
Because 5280 ft = 1 mi, divide by 5280.
$$3960\text{ ft} = \frac{3960}{5280} = \frac{3}{4}\text{ mi}$$

b. $3\frac{1}{2}$ lb to ounces

_____

c. 75 sec to minutes

_____

d. 380 min to hours

_____

Review these examples for Objective 3:
3.

    a. Convert 64 oz to pounds.

Use a unit fraction with pounds (the unit for your answer) in the numerator, and ounces (the unit being changed) in the denominator. Because 1 lb = 16 oz, the necessary unit fraction is

$$\frac{1\text{ lb}}{16\text{ oz}} \quad \begin{array}{l}\leftarrow \text{ Unit for your answer is pounds.}\\ \leftarrow \text{ Unit being changed is ounces.}\end{array}$$

Next, multiply 64 oz times this unit fraction.

Write 64 oz as the fraction $\dfrac{64\text{ oz}}{1}$ and divide out common units and factors wherever possible.

$$\frac{64\text{ oz}}{1} \cdot \frac{1\text{ lb}}{16\text{ oz}} = \frac{\overset{4}{\cancel{64}}\ \cancel{\text{oz}}}{1} \cdot \frac{1\text{ lb}}{\underset{1}{\cancel{16}}\ \cancel{\text{oz}}} = \frac{4 \cdot 1\text{ lb}}{1} = 4\text{ lb}$$

    b. Convert 8 lb to ounces.

Select the correct unit fraction to change 8 lb to ounces.

$$\frac{16\text{ oz}}{1\text{ lb}} \quad \begin{array}{l}\leftarrow \text{ Unit for your answer is ounces.}\\ \leftarrow \text{ Unit being changed is pounds.}\end{array}$$

Multiply 8 lb times the unit fraction.

$$\frac{8\text{ lb}}{1} \cdot \frac{16\text{ oz}}{1\text{ lb}} = \frac{8\ \cancel{\text{lb}}}{1} \cdot \frac{16\text{ oz}}{1\ \cancel{\text{lb}}} = \frac{8 \cdot 16\text{ oz}}{1} = 128\text{ oz}$$

Now Try:
3.

    a. Convert 12 yd to inches.

_____

    b. Convert 4 in. to feet.

_____

4. Convert using unit fractions.

a. Convert 6 qt to gallons.

First select the correct unit fraction.

$\dfrac{1\ \text{gal}}{4\ \text{qt}}$ ← Unit for your answer is gallons.
← Unit being changed is quarts.

Next multiply.

$$\frac{6\ \text{qt}}{1}\cdot\frac{1\ \text{gal}}{4\ \text{qt}}=\frac{\overset{3}{\cancel{6}}\ \cancel{\text{qt}}}{1}\cdot\frac{1\ \text{gal}}{\underset{2}{\cancel{4}}\ \cancel{\text{qt}}}=\frac{3}{2}\ \text{gal}=1\frac{1}{2}\ \text{gal}$$

b. Convert $6\frac{1}{2}$ lb to ounces.

Write $6\frac{1}{2}$ lb as an improper fraction.

$$\frac{6\frac{1}{2}\ \cancel{\text{lb}}}{1}\cdot\frac{16\ \text{oz}}{1\ \cancel{\text{lb}}}=\frac{13}{2}\cdot\frac{16}{1}\ \text{oz}$$

$$=\frac{13}{\underset{1}{\cancel{2}}}\cdot\frac{\overset{8}{\cancel{16}}}{1}\ \text{oz}$$

$$=104\ \text{oz}$$

c. Convert 340 hr to days.

$$\frac{\overset{85}{\cancel{340}}\ \cancel{\text{hr}}}{1}\cdot\frac{1\ \text{day}}{\underset{6}{\cancel{24}}\ \cancel{\text{hr}}}=\frac{85}{6}\ \text{day}=14\frac{1}{6}\ \text{day}$$

5. Sometimes you may need to use two or three unit fractions to complete a conversion.

a. Convert 110 pt to gallons.

Use the unit fraction $\dfrac{1\ \text{qt}}{2\ \text{pt}}$ to change pints to

quarts and the unit fraction $\dfrac{1\ \text{gal}}{4\ \text{qt}}$ to change

quarts to gallons. Notice how all the units divide out except gallons, which is the unit you want in the answer.

$$\frac{110\ \cancel{\text{pt}}}{1}\cdot\frac{1\ \cancel{\text{qt}}}{2\ \cancel{\text{pt}}}\cdot\frac{1\text{gal}}{4\ \cancel{\text{qt}}}=\frac{110}{8}\text{gal}=\frac{110\div2}{8\div2}\text{gal}=13\frac{3}{4}\text{gal}$$

13.75 gal is equivalent to $13\frac{3}{4}$ gal.

4. Convert using unit fractions.

a. Convert 4 mi to feet.

_____

b. Convert 3000 lb to tons.

_____

c. Convert $3\frac{1}{4}$ gal to quarts.

_____

5. Sometimes you may need to use two or three unit fractions to complete a conversion.
a. Convert 12 cups to gallons.

_____

b. Convert 3 miles to inches.

Use three unit fractions. The first one changes from miles to yards, the next one changes yards to feet, and the last one changes feet to inches. All the units divide out except inches, which is the unit you want in your answer.

$$\frac{3 \,\text{mi}}{1} \cdot \frac{1760 \,\text{yd}}{1 \,\text{mi}} \cdot \frac{3 \,\text{ft}}{1 \,\text{yd}} \cdot \frac{12 \,\text{in.}}{1 \,\text{ft}} = 190,080 \text{ inches}$$

b. Convert 5 days to minutes.

_____

Review these examples for Objective 4:

6. Answer each application problem.

a. Lee paid $4.65 for 14 oz of honey baked ham. What is the price per pound, to the nearest cent?

Step 1 Read the problem. The problem asks for the price per pound of the ham.

Step 2 Work out a plan. The weight of the ham is given in ounces, but the answer must be cost per pound. Convert ounces to pounds. The word per indicates division. You need to divide the cost by the number of pounds.

Step 3 Estimate a reasonable answer. To estimate, round $4.65 to $5, and round 14 ounces to 16 ounces. Then there are 16 oz in a pound, so there is about 1 pound. Finally, $5 \div 1 = \$5$ per pound is our estimate.

Step 4 Solve the problem. Use a unit fraction to convert 14 oz to pounds.

$$\frac{\overset{7}{\cancel{14}} \,\cancel{\text{oz}}}{1} \cdot \frac{1 \,\text{lb}}{\underset{8}{\cancel{16}} \,\cancel{\text{oz}}} = \frac{7}{8} \text{lb} = 0.875 \text{ lb}$$

Then divide to find the cost per pound.

$$\frac{\$4.65}{0.875} \approx \$5.31$$

Step 5 State the answer. The ham is $5.31 per pound (nearest cent).

Step 6 Check your work. The exact answer, $5.31 is close to the estimate of $5.

Now Try:

6. Answer each application problem.

a. Clarissa paid $1.79 for 4.5 oz of nuts. What is the cost per pound, to the nearest cent?

_____

b. At a preschool, each of 20 children drinks about $\frac{3}{4}$ c of juice with their snack each day. The school is open 5 days a week. How many quarts of juice are needed for 1 week of snacks?

Step 1 Read the problem. The problem asks for the number of quarts of juice needed for 1 week.

Step 2 Work out a plan. Multiply to find the number of cups of juice for one week. Then convert cups to quarts.

Step 3 Estimate a reasonable answer. To estimate, round $\frac{3}{4}$ cups to 1 cup. Then 1 cup times 20 children times 5 days is 100 cups. There are 4 cups in a quart. So, 100 cups $\div 4 = 25$ quarts is our estimate.

Step 4 Solve the problem. First multiply. Then use unit fractions to convert.

$$\frac{3}{4} \cdot 20 \cdot 5 = \frac{3}{\cancel{4}} \cdot \frac{\cancel{20}^{5}}{1} \cdot \frac{5}{1} = 75 \text{ cups}$$

$$\frac{75 \text{ cups}}{1} \cdot \frac{1 \text{ pt}}{2 \text{ cups}} \cdot \frac{1 \text{ qt}}{2 \text{ pt}} = \frac{75}{4} \text{ qt} = 18\frac{3}{4} \text{ qt}$$

Step 5 State the answer. The preschool needs $18\frac{3}{4}$ qt of juice for 1 week.

Step 6 Check your work. The exact answer of $18\frac{3}{4}$ qt is close to our estimate of 25 qt.

b. Sweet Suzie's Shop makes 49,000 lb of fudge each week. How many tons of fudge are produced each day of the 7-day week?

_____

Objective 1   **Learn the basic U.S. measurement units.**

For extra help, see Example 1 on page 478 of your text and Section Lecture video for Section 7.1.

Fill in the blanks.

  1.   1 T = _____ lb   **1.** _____

  2.   _____ qt = 1 gal   **2.** _____

3. 1 c = _____ fl oz   **3.** _____

Objective 2   Convert among U.S. measurement units using multiplication or division.

For extra help, see Example 2 on page 479 of your text and Section Lecture video for Section 7.1 and Exercise Solutions Clip 11.

Convert each measurement using multiplication or division.

    4.   12 ft to yards   **4.** _____

    5.   40 pt to gallons   **5.** _____

    6.   30 in to yards   **6.** _____

Objective 3   Convert among measurement units using unit fractions.

For extra help, see Examples 3–5 on pages 480–482 of your text and Section Lecture video for Section 7.1 and Exercise Solutions Clip 39.

Convert each measurement using unit fractions.

    7.   28 pt to gallons   **7.** _____

    8.   38 c to pints   **8.** _____

    9.   60 oz to pounds   **9.** _____

Objective 4   Solve application problems using U.S. measurement units.

For extra help, see Example 6 on pages 482–483 of your text and Section Lecture video for Section 7.1 and Exercise Solutions Clip 57.

Solve each application problem using the six problem-solving steps.

10.   Tony paid $2.84 for 6.5 oz of fudge. What is the cost   10. _____ per pound, to the nearest cent?

11.   At an office, each of 15 workers drinks about $1\frac{3}{4}$ c   11. _____

      of coffee with each day. The office is open 4 days a week. How many quarts of coffee are needed for a week?

Chapter 7 MEASUREMENT

7.2   The Metric System—Length

Learning Objectives
1   Learn the basic metric units of length.
2   Use unit fractions to convert among metric units.
3   Move the decimal point to convert among metric units.

Key Terms

Use the vocabulary terms listed below to complete each statement in exercises 1–3.

meter   prefix   metric conversion line

1.   Attaching a _____ such as "kilo-" or "milli" to the words "meter", "liter", or "gram" gives the names of larger or smaller units.

2.   A line showing the various metric measurement prefixes and their size relationship to each other is called a _____.

3.   The basic unit of length in the metric system is the _____.

Guided Examples

Review these examples for Objective 1:

1.   Write the most reasonable metric unit in each blank. Choose from km, m, cm, and mm.

   a. The man's foot is 28 _____ .

   28 cm because cm are used instead of inches.
   28 cm is about 11 inches.

   b. The living room is 4 _____ .

   4 m because m are used instead of feet.
   4 m is about 13 feet.

   c. Sam rode his bike 8 _____ on the trail.

   8 km because km are used instead of miles.
   8 km is about 5 miles.

Now Try:

1.   Write the most reasonable metric unit in each blank. Choose from km, m, cm, and mm.
   a. A strand of hair is 1 ____ wide.

   _____

   b. The truck sped down the highway at 120 _____ per hour.

   _____

   c. The width of a piece of paper is 21.6 _____ .

   _____

Review these examples for Objective 2:

2. Convert each measurement using unit fractions.

  a. 8 m to cm

Put the unit for the answer (cm) in the numerator of the unit fraction; put the unit you want to change (m) in the denominator.

$\dfrac{100 \text{ cm}}{1 \text{ m}}$ ← Unit for answer
← Unit being changed

Multiply. Divide out common units where possible.

$$8\text{m}\cdot\frac{100\text{cm}}{1\text{m}} = \frac{8\cancel{\text{m}}}{1}\cdot\frac{100\text{cm}}{1\cancel{\text{m}}} = \frac{8\cdot100\text{cm}}{1} = 800\text{cm}$$

8 m = 800 cm

  b. 13.5 mm to cm

Multiply by a unit fraction that allows you to divide out millimeters.

$$\frac{13.5\ \cancel{\text{mm}}}{1}\cdot\frac{1\text{ cm}}{10\ \cancel{\text{mm}}} = \frac{13.5\text{ cm}}{10} = 1.35 \text{ mm}$$

13.5 mm = 1.35 cm
There are 10 mm in a cm, so 13.5 mm will be a smaller part of a cm.

1000	100	10	1	$\frac{1}{10}$	$\frac{1}{100}$	$\frac{1}{1000}$
km	hm	dam	m	dm	cm	mm

Review these examples for Objective 3:

3. Use the metric conversion line to make the following conversions.

  a. 62.892 km to m

Find km on the metric conversion line. To get m, you move three places to the right. So move the decimal point in 62.892 three places to the right.
    62.892 km = 62,892 m

  b. 47.6 cm to m

    Find cm on the conversion line. To get m, move two places to the left. So move the decimal point two places to the left.
        47.6 cm = 0.476 m

  c. 52.8 mm to cm

From mm to cm is one place to the left.
    52.8 mm = 5.28 cm

Now Try:

2. Convert each measurement using unit fractions.

  a. 5400 m to km

_____

  b. 7.6 cm to mm

_____

Now Try:

3. Use the metric conversion line to make the following conversions.

  a. 67.5 cm to mm

_____

  b. 986.5 m to km

_____

  c. 4.31 m to cm

_____

Name: Date:
Instructor: Section:

4. Convert using the metric conversion line.

   a. 1.79 m to mm

Moving from m to mm is going three places to the right. In order to move the decimal point in 1.79 three places to the right, you must add a 0 as a place holder.
   1.790
1.79 m = 1790 mm

   b. 50 cm to m

From cm to m is two places to the left. The decimal point in 50 starts at the far right side because 50 is a whole number. Then move it two places to the left.

50 cm = 0.50 m and 0.50 m is equivalent to 0.5 m.

   c. 19 m to km

From m to km is three places to the left. The decimal point in 19 starts at the far right side. In order to move it three places to the left, you must write in one zero as a placeholder.

19 m = 0.019 km

4. Convert using the metric conversion line.
   a. 23 m to mm

   _____

   b. 4.2 cm to m

   _____

   c. 950 m to km

   _____

Objective 1   Learn the basic metric units of length.

For extra help, see Example 1 on page 489 of your text and Section Lecture video for Section 7.2 and Exercise Solutions Clip 13, 15, 17, and 19.

Choose the most reasonable metric unit. Choose from **km, m, cm,** *or* **mm**.

  1.   the width of a twin bed   **1.** _____

  2.   the thickness of a dime   **2.** _____

  3.   the distance driven in 2 hours   **3.** _____

Objective 2   Use unit fractions to convert among metric units.

For extra help, see Example 2 on page 490 of your text and Section Lecture video for Section 7.2 and Exercise Solutions Clip 29, 31, and 35.

Convert each measurement using unit fractions.

4.   25.87 m to centimeters   4. _____

5.   450 m to kilometers   5. _____

6.   140 millimeters to meters   6. _____

Objective 3   Move the decimal point to convert among metric units.

For extra help, see Examples 3–4 on pages 491–492 of your text and Section Lecture video for Section 7.2 and Exercise Solutions Clip 29, 31, and 35.

Convert each measurement using the metric conversion line.

$$1000 \quad 100 \quad 10 \quad 1 \quad \tfrac{1}{10} \quad \tfrac{1}{100} \quad \tfrac{1}{1000}$$

km   hm   dam   m   dm   cm   mm

7.   1.94 cm to millimeters   7. _____

8.   10.35 km to meters   8. _____

9.   3.5 cm to kilometers   9. _____

Chapter 7 MEASUREMENT

7.3   The Metric System—Capacity and Weight (Mass)

Learning Objectives	
1	Learn the basic metric units of capacity.
2	Convert among metric capacity units.
3	Learn the basic metric units of weight (mass).
4	Convert among metric weight (mass) units.
5	Distinguish among basic metric units of length, capacity, and weight (mass).

Key Terms

Use the vocabulary terms listed below to complete each statement in exercises 1–2.

       liter   **gram**

1. The basic unit of weight (mass) in the metric system is the _____.

2. The basic unit of capacity in the metric system is the _____.

Guided Examples

Review these examples for Objective 1:

1. Write the most reasonable metric unit in each blank. Choose from L and mL.

   a. The bottle of sunscreen held 118 _____ .

118 mL because 118 L would be about 118 qts, which is too much.

   b. The soda bottle has 2 _____ .

2 L because 2 mL would be less than a teaspoon.

Now Try:

1. Write the most reasonable metric unit in each blank. Choose from L and mL.

   a. Ty added 5 _____ to the tea.

      _____

   b. The water tank holds 100 _____ .

      _____

Review these examples for Objective 2:

2. Convert using the metric conversion line or unit fractions.

   a. 5.4 L to mL

Using the metric conversion line:
From L to mL is three places to the right.
   5.4 L = 5400 mL

Using unit fractions:
Multiply by a unit fraction that allows you to divide out liters.

$$\frac{5.4 \; \cancel{L}}{1} \cdot \frac{1000 \; mL}{1 \; \cancel{L}} = 5400 \; mL$$

Now Try:

2. Convert using the metric conversion line or unit fractions.

   a. 973 mL to L

      _____

b. 92 mL to L

Using the metric conversion line:
From mL to L is three places to the left.
    92 mL = 0.092 L

Using unit fractions:
Multiply by a unit fraction that allows you to
divide out mL.

$$\frac{9.2 \ mL}{1} \cdot \frac{1 \ L}{1000 \ mL} = 0.092 \ L$$

b. 3.85 L to mL

_____

Review these examples for Objective 3:
3. Write the most reasonable metric unit in each
    blank. Choose from kg, g, and mg.

    a. The watermelon weighed 18 _____ .

    18 kg because kilograms are used instead of
    pounds.
    18 kg is about 40 pounds.

    b. The vitamin tablet was 400 _____ .

    400 mg because 400 g would be more than two
    hamburgers, which is too much.

    c. A birthday card weighs 8 _____ .

    8 grams because 8 kg would be too heavy.

Now Try:
3. Write the most reasonable
    metric unit in each blank.
    Choose from kg, g, and mg.
    a. A fashion model weighs
    48 _____ .

    _____

    b. Jacob's football weighs
    590 _____ .

    _____

    c. The pencil lead weighs
    3 _____ .

    _____

Review these examples for Objective 4:
4. Convert using the metric conversion line or unit
    fractions.

    a. 26 mg to g

    Using the metric conversion line:
    From mg to g is three places to the left.
        26 mg = 0.026 g

    Using unit fractions:
    Multiply by a unit fraction that allows you to
    divide out mg.

$$\frac{26 \ mg}{1} \cdot \frac{1 \ g}{1000 \ mg} = 0.026 \ g$$

Now Try:
4. Convert using the metric
    conversion line or unit fractions.

    a. 3.72 g to mg

    _____

b. 97.34 kg to g

Using the metric conversion line:
From kg to g is three places to the right.
   97.34 kg = 97,340 g

Using unit fractions:
Multiply by a unit fraction that allows you to divide out kg.

$$\frac{97.34\ \cancel{kg}}{1} \cdot \frac{1000\ g}{1\ \cancel{kg}} = 97,340\ g$$

b. 84 g to kg

_____

Review these examples for Objective 5:

5. First decide which type of unit is needed: length, capacity, or weight. Then write the most appropriate metric unit in the blank. Choose from km, m, cm, mm, L, mL, kg, g, and mg.

  a. The tree is 9 _____ high.

Use length units because of the word high.
The tree is 9 m high.

  b. A ten-carat diamond weighs 2 _____ .

Use weight units because of the word weigh.
The ten-carat diamond weighs 2 g.

  c. The milk container has 3.78 _____ .

Use capacity units because milk is a liquid.
The milk container has 3.78 L.

Now Try:

5. First decide which type of unit is needed: length, capacity, or weight. Then write the most appropriate metric unit in the blank. Choose from km, m, cm, mm, L, mL, kg, g, and mg.
  a. The moisturizer jar has 125 _____ .

_____

  b. This is a 1.19 _____ box of breakfast cereal.

_____

  c. The boy walked 400 _____ to the bus stop.

_____

Objective 1 Learn the basic metric units of capacity.

For extra help, see Example 1 on page 496 of your text and Section Lecture video for Section 7.3 and Exercise Solutions Clip 11.

Choose the most reasonable metric unit. Choose from **L**, *or* **ml**.

1.   the amount of soda in a can   1. _____

2.   the amount of orange juice in a large bottle   2. _____

3.   the amount of cough syrup in one dose   3. _____

Objective 2 Convert among metric capacity units.

For extra help, see Example 2 on pages 496–497 of your text and Section Lecture video for Section 7.3.

Convert each measurement. Use unit fractions or the metric conversion line.

4.   2.5 L to milliliters   **4.** _____

5.   836 kL to liters   **5.** _____

6.   7863 mL to liters   **6.** _____

Objective 3 Learn the basic metric units of weight (mass).

For extra help, see Example 3 on page 498 of your text and Section Lecture video for Section 7.3 and Exercise Solutions Clip 7 and 13.

Choose the most reasonable metric unit. Choose from **kg**, **g**, *or* **mg**.

7.   the weight (mass) of a vitamin pill   **7.** _____

8.   the weight (mass) of a car   **8.** _____

9.   the weight (mass) of an egg   **9.** _____

Objective 4 Convert among metric weight (mass) units.

For extra help, see Example 4 on page 499 of your text and Section Lecture video for Section 7.3 and Exercise Solutions Clip 39, 41, and 43.

Convert each measurement. Use unit fractions or the metric conversion line.

10.   27,000 g to kilograms   **10.** _____

11.   0.76 kg to grams   **11.** _____

12.   4.7 g to milligrams   **12.** _____

Objective 5   **Distinguish among basic metric units of length, capacity, and weight (mass).**

For extra help, see Example 5 on page 500 of your text and Section Lecture video for Section 7.3 and Exercise Solutions Clip 57 and 61.

Choose the most reasonable metric unit. Choose from **km**, **m**, **cm**, **mm**, **mL**, **L**, **kg**, **g**, *or* **mg**.

13.   Buy a 5 _____ bottle of water.   **13.** _____

14.   The piece of wood weighs 5 _____.   **14.** _____

15.   A paperclip is 3 _____ long.   **15.** _____

Chapter 7 MEASUREMENT

7.4   Problem Solving with Metric Measurement

Learning Objectives
1   Solve application problems involving metric measurements.

Key Terms

Use the vocabulary terms listed below to complete each statement in exercises 1–3.

meter   liter   gram

1.   A _____ is the weight of 1 mL of water.

2.   A _____ is a little longer than a yard.

3.   A _____ is a little more than one quart.

Guided Examples

Review these examples for Objective 1:

1.   Colby cheese is on sale at $9.89 per kilogram. Nelson bought 750 g of the cheese. How much did he pay, to the nearest cent?

Step 1 Read the problem. The problem asks for the cost of 750 g of cheese.

Step 2 Work out a plan. The price is $9.89 per kilogram, but the amount Nelson bought is given in grams. Convert grams to kilograms. Then multiply the weight by the cost per kilogram.

Step 3 Estimate a reasonable answer. Round the cost of 1 kg from $9.89 to $10. There are 1000 g in a kilogram, so 750 g is about $\frac{3}{4}$ of a kilogram.

Nelson buys about $\frac{3}{4}$ of a kilogram, so $\frac{3}{4}$ of $10 = $7.50 is our estimate.

Step 4 Solve the problem. Use a unit fraction to convert 750 g to kilograms.

$$\frac{750 \not{g}}{1} \cdot \frac{1 \text{ kg}}{1000 \not{g}} = 0.75 \text{ kg}$$

Now multiply 0.75 kg times the cost per kilogram.

Now Try:

1.   A ball of yarn weighs 140 g. Ellie knitted a sweater for her boyfriend that used 11 balls of yarn. How many kilograms did the finished sweater weigh?

_____

$$\frac{0.75 \cancel{kg}}{1} \cdot \frac{\$9.89}{1 \cancel{kg}} = \$7.4175 \approx \$7.42$$

Step 5 State the answer. Nelson paid $7.42, rounded to the nearest cent.

Step 6 Check your work. The exact answer of $7.42 is close to our estimate of $7.50.

2. A 70-L drum is filled with oil which is to be packaged into 140-mL bottles. How many bottles can be filled.?

Step 1 Read the problem. The problem asks for the number of filled bottles.

Step 2 Work out a plan. The given amount is in liters, but the capacity of the bottles is in mL. Convert liters to milliliters, then divide by 140 mL (the capacity of the bottles).

Step 3 Estimate a reasonable answer. To estimate round 140 mL to 100 mL. Then 70 L = 70,000 mL, and $70,000 \div 100 = 700$ bottles.

Step 4 Solve the problem. On the metric conversion line, moving from L to mL is three places to the right, so move the decimal point in 70 L three places to the right. Then divide by 140.

70 L = 70,000 mL

$$\frac{70,000 \text{ mL}}{140 \text{ mL}} = 500 \text{ bottles}$$

Step 5 State the answer. 500 bottles will be filled.

Step 6 Check your work. The exact answer of 500 bottles is close to our estimate of 700 bottles.

3. Lucy purchased 1 m 60 cm of fabric at $6.25 per meter for a jacket and 1m 80 cm of fabric at $4.75 per m for a dress. How much did she spend in total?

Step 1 Read the problem. Lucy purchased two quantities of two different kinds of fabric. Find the total she spent on fabric.

Step 2 Work out a plan. The lengths involve two

2. If 1.8 kg of candy is to be divided equally among 9 children, how many grams will each child receive?

_____

3. A fish tank can hold up to 75.6 L of water. If there are 70,000 mL of water in the tank, how many more milliliters of water can the tank hold?

_____

Name: Date:
Instructor: Section:

units, m and cm. Rewrite both lengths in meters, determine the price, and find the total.

Step 3 Estimate a reasonable answer. To estimate, 1 m 60 cm can be rounded to 2 m. Round 1 m 80 cm to 2 m. Also, $6.25 can be rounded to $6 and $4.75 can be rounded to $5. Then, $2 \times \$6 + 2 \times \$5 = \$12 + \$10 = \$22$, our estimate.

Step 4 Solve the problem. Rewrite the lengths in meters.

1 m →	1.0 m	1 m →	1.0 m
plus 60 cm →	+ 0.6 m	plus 80 cm →	+ 0.8 m
	1.6 m		1.8 m

Jacket: $1.6 \times \$6.25 = \$10$
Dress: $1.8 \times \$4.75 = \underline{\$\ 8.55}$
                                        $18.55 total

Step 5 State the answer. The total spent is $18.55.

Step 6 Check your work. The exact answer of $18.55 is close to our estimate of $22.

Objective 1   Solve application problems involving metric measurements.

For extra help, see Examples 1–3 on pages 507–508 of your text and Section Lecture video for Section 7.4 and Exercise Solutions Clip 3, 9, 11, and 13.

Solve each application problem. Round money answers to the nearest cent.

1. Metal chain costs $5.26 per meter. Find the cost of 2 m 47 cm of the chain. Round your answer to the nearest cent.

1. _____

2. Henry bowls with a 7-kg bowling ball, while Denise uses a ball that weighs 5 kg 750g. How much heavier is Henry's bowling ball than Denise's?

2. _____

3. The label on a bottle of pills says that there are 3.5 mg of the medication in 5 pills. If a patient needs to take 8.4 mg of the medication, how many pills does he need to take?

3. _____

Chapter 7 MEASUREMENT

7.5   Metric–U.S. Measurement Conversions and Temperature

Learning Objectives
1   Use unit fractions to convert between metric and U.S. measurement units.
2   Learn common temperatures on the Celsius scale.
3   Use formulas to convert between Celsius and Fahrenheit temperatures.

Key Terms

Use the vocabulary terms listed below to complete each statement in exercises 1–2.

Celsius   **Fahrenheit**

1.   The _____ scale is used to measure temperature in the metric system.

2.   The _____ scale is used to measure temperature in the U.S. customary system.

Guided Examples

Review these example for Objective 1:
1.   Convert from 16 in. to centimeters.

We're changing from a U.S length unit to a metric length unit. In the "U.S. to Metric Units" side of the table, you see that 1 inch ≈ 2.54 centimeters. Two unit fractions can be written using that information.

$$\frac{1 \text{ in.}}{2.54 \text{ cm}} \quad \text{or} \quad \frac{2.54 \text{ cm}}{1 \text{ in.}}$$

Multiply by the unit fraction that allows you to divide out inches.

$$16 \text{ in.} \cdot \frac{2.54 \text{ cm}}{1 \text{ in.}} = \frac{16 \text{ in.}}{1} \cdot \frac{2.54 \text{ cm}}{1 \text{ in.}} = 40.64 \text{ cm}$$

16 in. ≈ 40.64 cm

Now Try:
1.   Convert from 12.9 m to feet.

_____

2. Convert using unit fractions. Round your answers to the nearest tenth.

a. 22.5 pounds to kilograms

Look in the "U.S. to Metric Units" side of the table to see that 1 pound ≈ 0.45 kilograms. Use this information to write a unit fraction that allows you to divide out pounds.

$$\frac{22.5 \text{ lb}}{1} \cdot \frac{0.45 \text{ kg}}{1 \text{ lb}} = 10.125 \text{ kg}$$

22.5 lb ≈ 10.1 kg

b. 46.8 L to quarts

Look in the "Metric to U.S. Units" side of the table to see that $1 \text{ L} \approx 1.06$ quarts. Write a unit fraction that allows you to divide out liters.

$$\frac{46.8 \text{ L}}{1} \cdot \frac{1.06 \text{ qt}}{1 \text{ L}} = 49.608 \text{ qt}$$

46.8 L ≈ 49.6 qt

2. Convert using unit fractions. Round your answers to the nearest tenth.

a. 9.68 kg to pounds

_____

b. 20 gallons to liters

_____

Review these examples for Objective 2:

3. Choose the metric temperature that is most reasonable for each situation.

a. hot water in a bathtub
   27°C   40°C   100°C

27°C is too cold and 100°C is too hot.
40°C is reasonable.

b. fall day
   13°C   50°C   65°C

50°C and 65°C are too hot.
13°C is reasonable.

Now Try:

3. Choose the metric temperature that is most reasonable for each situation.

a. hot cocoa
   10°C   35°C   65°C

_____

b. ice cream
   –10°C   4°C   30°C

_____

Review these examples for Objective 3:

4. Convert 77°F to Celsius.

Use the formula and follow the order of operations.

$$C = \frac{5(F-32)}{9}$$

$$= \frac{5(77-32)}{9}$$

$$= \frac{5(45)}{9}$$

$$= \frac{5(\overset{5}{\cancel{45}})}{\underset{1}{\cancel{9}}}$$

$$= 25$$

Thus, 77°F = 25°C.

5. Convert 30°C to Fahrenheit.

Use the formula and follow the order of operations.

$$F = \frac{9 \cdot C}{5} + 32$$

$$= \frac{9 \cdot 30}{5} + 32$$

$$= \frac{9 \cdot \overset{6}{\cancel{30}}}{\underset{1}{\cancel{5}}} + 32$$

$$= 54 + 32$$

$$= 86$$

Thus, 30°C = 86°F.

Now Try:

4. Convert 122°F to Celsius.

_____

5. Convert 150°C to Fahrenheit.

_____

Objective 1   Use unit fractions to convert between metric and U.S. measurement units.

For extra help, see Examples 1–2 on pages 511–512 of your text and Section Lecture video for Section 7.5 and Exercise Solutions Clip 7, 9, and 13.

Use the table in your textbook and unit fractions to make the following conversions. Round answers to the nearest tenth.

1. 291 mi to kilometers

1. _____

2. 7 L to gallons

2. _____

3.   26 oz to grams   **3.** _____

Objective 2   Learn common temperatures on the Celsius scale.

For extra help, see Example 3 on page 513 of your text and Section Lecture video for Section 7.5.

Choose the most reasonable temperature for each situation.

4.   hot coffee   **4.** _____
          35°C   60°C   100°C

5.   Normal body temperature   **5.** _____
          37°C   37°F

6.   Oven temperature   **6.** _____
          300°C   300°F

Objective 3   Use formulas to convert between Celsius and Fahrenheit
**   temperatures.**

For extra help, see Examples 4–5 on page 514 of your text and Section Lecture video for Section 7.5 and Exercise Solutions Clip 29 and 31.

Use the conversions formulas and the order of operations to convert Fahrenheit temperatures to Celsius and Celsius temperatures to Fahrenheit. Round your answers to the nearest degree, if necessary.

7.   62°F   **7.** _____

8.   10°C   **8.** _____

Solve the application problem. Round to the nearest degree, if necessary.

9.   A recipe for roast beef calls for an oven temperature   **9.** _____
          of 400°F. What is the temperature in degrees
          Celsius?

Name: Date:
Instructor: Section:

Chapter 8 GEOMETRY

8.1   Basic Geometric Terms

Learning Objectives
1   Identify and name lines, line segments, and rays.
2   Identify parallel and intersecting lines.
3   Identify and name angles.
4   Classify angles as right, acute, straight, or obtuse.
5   Identify perpendicular lines.

Key Terms

Use the vocabulary terms listed below to complete each statement in exercises 1–13.

point   line   line segment   ray

angle   degrees   right angle   acute angle

obtuse angle   straight angle

intersecting lines   perpendicular lines   parallel lines

1.   A _____ is a part of a line that has one endpoint and which extends infinitely in one direction.

2.   Two lines that intersect to form a right angle are _____.

3.   An angle whose measure is between 90° and 180° is an _____.

4.   A _____ is a location in space.

5.   Two rays with a common endpoint form an _____.

6.   A set of points that form a straight path that extends infinitely in both directions is called a _____.

7.   An angle that measures less than 90° is called an _____.

8.   Angles are measured using _____.

9.   An angle whose measure is exactly 180° is called a _____.

10.   Two lines in the same plane that never intersect are _____.

11.   A part of a line with two endpoints is a _____.

12.   Two lines that cross at one point are _____.

13.   An angle whose measure is exactly 90° is a _____.

Name: Date:
Instructor: Section:

Guided Examples

Review these examples for Objective 1:

1. Identify each figure below as a line, line segment, or ray, and name it using the appropriate symbol.

a.

This figure has two endpoints, so it is a line segment named $\overline{GH}$ or $\overline{HG}$.

b.

This figure starts at point B and goes on forever in one direction, so it is a ray named $\overrightarrow{BA}$.

c.

This figure goes on forever in both directions, so it is a line named $\overleftrightarrow{PQ}$ or $\overleftrightarrow{QP}$.

Review these examples for Objective 2:

2. Label each pair of the lines as appearing to be parallel or as intersecting.

a.

The lines in this figure cross, so they are intersecting lines.

Now Try:

1. Identify each figure below as a line, line segment, or ray, and name it using the appropriate symbol.

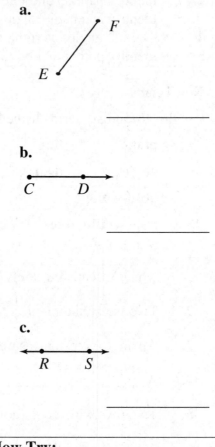

a.

_____

b.

_____

c.

_____

Now Try:

2. Label each pair of the lines as appearing to be parallel or as intersecting.

a.

_____

b.

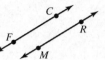

The lines in this figure do not intersect; they appear to be parallel.

b.

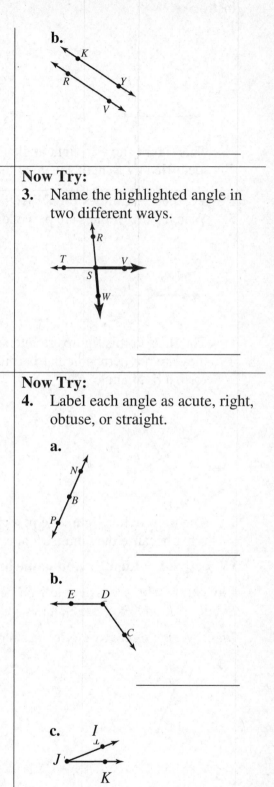

_____

Review this example for Objective 3:

3. Name the highlighted angle in two different ways.

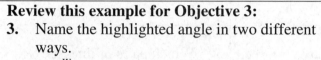

The angle can be named $\angle ZRM$ or $\angle MRZ$. It cannot be named $\angle R$, using the vertex alone, because four different angles have R as their vertex.

Now Try:

3. Name the highlighted angle in two different ways.

_____

Review these examples for Objective 4:

4. Label each angle as acute, right, obtuse, or straight.

a.

This figure shows a straight angle (exactly 180°).

b.

This figure shows an obtuse angle (more than 90° but less than 180°).

c.

This figure shows an acute angle (less than 90°).

Now Try:

4. Label each angle as acute, right, obtuse, or straight.

a.

_____

b.

_____

c.

_____

d.

This figure shows a right angle (exactly 90° and identified by a small square at the vertex).

d.

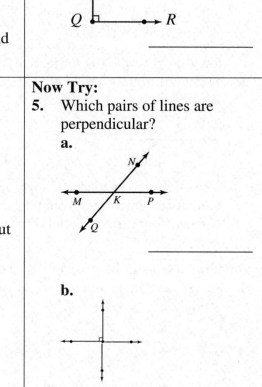

_____

Review these examples for Objective 5:

5. Which pairs of lines are perpendicular?

a.

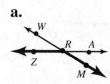

The lines in this figure are intersecting lines, but they are not perpendicular because they do not form a right angle.

b.

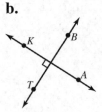

The lines in this figure are perpendicular to each other because they intersect at right angles.

Now Try:

5. Which pairs of lines are perpendicular?

a.

_____

b.

_____

Objective 1   Identify and name lines, line segments, and rays.

For extra help, see Example 1 on page 534 of your text and Section Lecture video for Section 8.1 and Exercise Solutions Clip 3 and 7.

Identify each figure as a line, line segment, or ray, and name it.

1.

1. _____

2.

2. _____

3.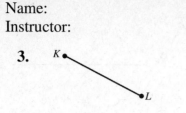

3. _____

Objective 2   Identify parallel and intersecting lines.

For extra help, see Example 2 on page 535 of your text and Section Lecture video for Section 8.1 and Exercise Solutions Clip 11.

Label each pair of lines as appearing to be **parallel** *or* **intersecting**.

4.

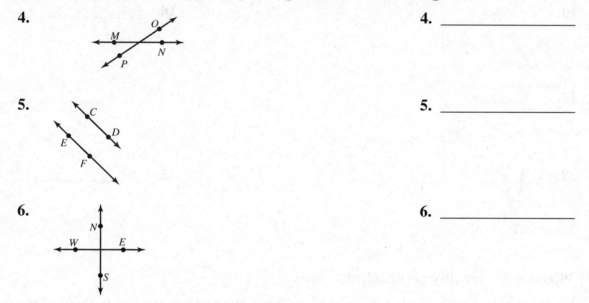

4. _____

5.

5. _____

6.

6. _____

Objective 3   Identify and name angles.

For extra help, see Example 3 on page 536 of your text and Section Lecture video for Section 8.1 and Exercise Solutions Clip 15.

Name each angle drawn with darker rays by using the three-letter form of identification.

7.

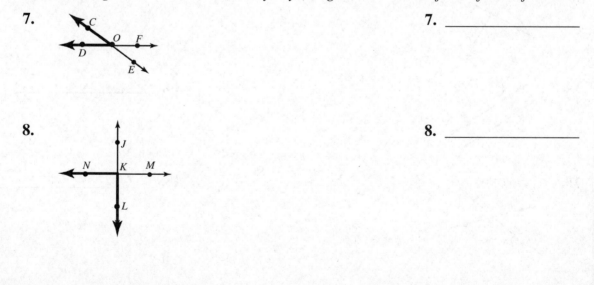

7. _____

8.

8. _____

9.

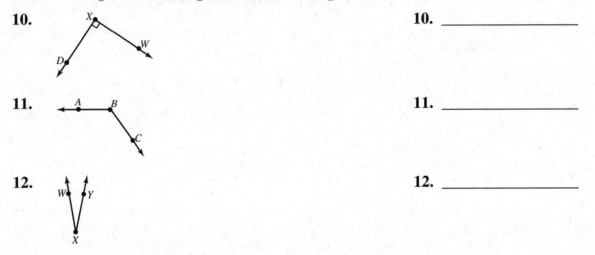

9. _____

Objective 4   Classify angles as right, acute, straight, or obtuse.

For extra help, see Example 4 on page 537 of your text and Section Lecture video for Section 8.1 and Exercise Solutions Clip 19, 21, and 23.

Label each angle as **acute**, **right**, **obtuse**, *or* **straight**.

10.

10. _____

11.

11. _____

12.

12. _____

Objective 5   Identify perpendicular lines.

For extra help, see Example 5 on page 538 of your text and Section Lecture video for Section 8.1 and Exercise Solutions Clip 13.

Label each pair of lines as appearing to be **parallel**, **perpendicular**, *or* **intersecting**.

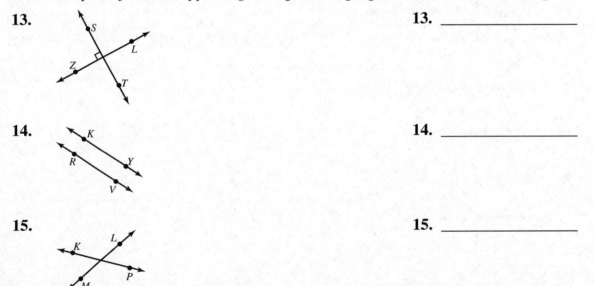

13.

13. _____

14.

14. _____

15.

15. _____

Chapter 8 GEOMETRY

8.2   Angles and Their Relationships

Learning Objectives

1   Identify complementary angles and supplementary angles and find the *measure* of a complement or supplement of a given angle.

2   Identify congruent angles and vertical angles and use this knowledge to find the measures of angles.

Key Terms

Use the vocabulary terms listed below to complete each statement in exercises 1–4.

complementary angles   supplementary angles

congruent angles   vertical angles

1.   The nonadjacent angles formed by two intersecting lines are called

_____.

2.   Angles whose measures are equal are called _____.

3.   Two angles whose measures sum to 180° are _____.

4.   Two angles whose measures sum to 90° are _____.

Guided Examples

Review these examples for Objective 1:

1.   Identify each pair of complementary angles.

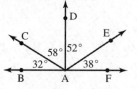

∠*BAC* (32°) and ∠*CAD* (58°) are complementary angles because
$$32° + 58° = 90°$$
∠*DAE* (52°) and ∠*EAF* (38°) are complementary angles because
$$52° + 38° = 90°$$

2.   Find the complement of each angle.

a. 66°

Find the complement of 66° by subtracting.
$$90° - 66° = 24°$$

Now Try:

1.   Identify each pair of complementary angles.

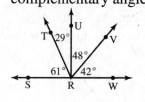

_____

2.   Find the complement of each angle.

a. 72°

_____

b. 4°

Find the complement of 4° by subtracting.

$$90° - 4° = 86°$$

3. In the figures below, $\angle ABC$ measures 150°, $\angle WXY$ measures 30°, $\angle MKQ$ measures 30°, and $\angle QKP$ measures 150°. Identify each pair of supplementary angles.

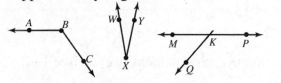

$\angle ABC$ and $\angle WXY$, because $150° + 30° = 180°$

$\angle MKQ$ and $\angle QKP$, because $30° + 150° = 180°$

$\angle ABC$ and $\angle MKQ$, because $150° + 30° = 180°$

$\angle WXY$ and $\angle QKP$, because $30° + 150° = 180°$

4. Find the supplement of each angle.

a. 38°

Find the supplement of 38° by subtracting.

$$180° - 38° = 142°$$

b. 121°

Find the supplement of 121° by subtracting.

$$180° - 121° = 59°$$

Review this example for Objective 2:

5. Identify the angles that are congruent.

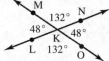

Congruent angles measure the same number of degrees.

$\angle LKM \cong \angle OKN$ and $\angle LKO \cong \angle MKN$.

---

b. 12°

_____

3. In the figures below, $\angle OQP$ measures 35°, $\angle PQN$ measures 145°, $\angle RST$ measures 35°, and $\angle BMC$ measures 145°. Identify each pair of supplementary angles.

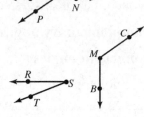

_____

4. Find the supplement of each angle.

a. 82°

_____

b. 168°

_____

Now Try:

5. Identify the angles that are congruent.

_____

6. Identify the vertical angles in this figure.

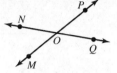

∠*POQ* and ∠*NOM* are vertical angles because they do not share a common side and they are formed by two intersecting lines ($\overleftrightarrow{NQ}$ and $\overleftrightarrow{MP}$).

∠*NOP* and ∠*MOQ* are also vertical angles.

7. In the figure below, ∠*APB* measures 93° and ∠*BPC* measures 37°. Find the measures of the indicated angles.

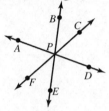

a. ∠*EPF*

∠*EPF* and ∠*BPC* are vertical angles, so they are congruent. This means they measure the same number of degrees.
The measure of ∠*BPC* is 37°, so the measure of ∠*EPF* is 37° also.

b. ∠*DPE*

∠*DPE* and ∠*APB* are vertical angles, so they are congruent.
The measure of ∠*APB* is 93°, so the measure of ∠*DPE* is 93° also.

c. ∠*CPD*

Look at ∠*CPD*, ∠*BPC*, and ∠*APB*. Notice that $\overrightarrow{PA}$ and $\overrightarrow{PD}$ go in opposite directions. Therefore, ∠*APD* is a straight angle and measures 180°. To find the measure of ∠*CPD*, subtract the sum of the other two angles from 180°.

$$180° - (93° + 37°) = 180° - (130°) = 50°$$

The measure of ∠*CPD* is 50°.

6. Identify the vertical angles in this figure.

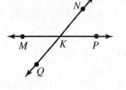

_____

7. In the figure below, ∠*AGF* measures 33° and ∠*BGC* measures 105°. Find the measures of the indicated angles.

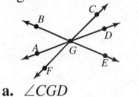

a. ∠*CGD*

_____

b. ∠*EGF*

_____

c. ∠*DGE*

_____

d. ∠*FPA*

∠*FPA* and ∠*CPD* are vertical angles, so they are congruent. We know from part (c) that the measure of ∠*CPD* is 50°, so the measure of ∠*FPA* is 50° also.

d. ∠*BGA*

_____

Objective 1 **Identify complementary angles and supplementary angles and find the *measure* of a complement or supplement of a given angle.**

For extra help, see Examples 1–4 on pages 541–542 of your text and Section Lecture video for Section 8.2 and Exercise Solutions Clip 3 and 5.

Find the complement of the angle.

1. 43°

1. _____

Find the supplement of the angle.

2. 16°

2. _____

Identify each pair of supplementary angles.

3.

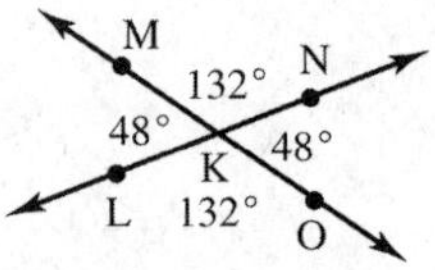

3. _____

Objective 2 **Identify congruent angles and vertical angles and use this knowledge to find the measures of angles.**

For extra help, see Examples 5–7 on pages 542–543 of your text and Section Lecture video for Section 8.2 and Exercise Solutions Clip 17 and 19.

In the figure below, identify the angles that are congruent.

4.

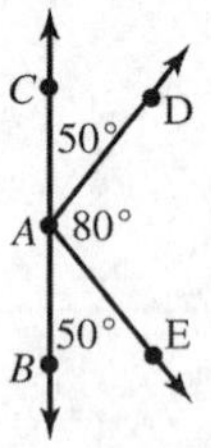

4. _____

In the figure below, identify all the vertical angles.

5.

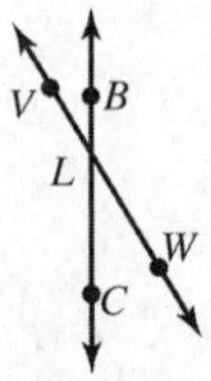

5. _____

In the figure below, $\angle ABE$ *measures 73° and* $\angle FEB$ *measures 107°. Find the measures of the indicated angles.*

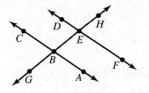

6. $\angle CBG$ and $\angle DEH$ **6.** _____

Chapter 8 GEOMETRY

8.3 Rectangles and Squares

Learning Objectives
1   Find the perimeter and area of a rectangle.
2   Find the perimeter and area of a square.
3   Find the perimeter and area of a composite figure.

Key Terms

Use the vocabulary terms listed below to complete each statement in exercises 1–4.

     perimeter   **area**   **rectangle**   **square**

1. The number of square units in a region is called the _____ of the region.

2. A four-sided figure with four right angles is called a _____.

3. The distance around the outside edges of a figure is called the _____ of the figure.

4. A rectangle with four equal sides is called a _____.

Guided Examples

Review these examples for Objective 1:

1. Find the perimeter of a rectangle that is 4 cm by 8 cm.

   You can use the formula, as shown below.
   $$P = 2 \cdot l \;\; + 2 \cdot w$$
   $$P = 2 \cdot 4 \text{ cm} + 2 \cdot 8 \text{ cm}$$
   $$P = \;\; 8 \text{ cm} \; + \;\; 16 \text{ cm}$$
   $$P = 24 \text{ cm}$$

   Or, you can add up the lengths of the four sides.
   $$P = 4 \text{ cm} + 4 \text{ cm} + 8 \text{ cm} + 8 \text{ cm}$$
   $$P = 24 \text{ cm}$$

   Either method will give you the correct result.

Now Try:

1. Find the perimeter of a rectangle that is 6 ft by 5 ft.

   _____

2. Find the area of a rectangle that is 4 cm by 8 cm.

The length is 8 cm (the longer measurement) and the width is 4 cm. Then use the formula for the area of a rectangle, $A = l \cdot w$.

$A = l \cdot w$

$A = 8 \text{ cm} \cdot 4 \text{ cm}$

$A = 32 \text{ cm}^2$

The area of the rectangle is 32 cm^2.

2. Find the area of a rectangle that is 6 ft by 5 ft.

_____

Review these examples for Objective 2:

3.

a. Find the perimeter of a square where each side measures 13 ft.

Use the formula.

$P = 4 \cdot s$

$P = 4 \cdot 13 \text{ ft}$

$P = 52 \text{ ft}$

Or add up the four sides.

$P = 13 \text{ ft} + 13 \text{ ft} + 13 \text{ ft} + 13 \text{ ft}$

$P = 52 \text{ ft}$

b. Find the area of a square where each side measures 13 ft.

Use the formula for area of a square.

$A = s^2$

$A = s \cdot s$

$A = 13 \text{ ft} \cdot 13 \text{ ft}$

$A = 169 \text{ ft}^2$

Now Try:

3.

a. Find the perimeter of a square where each side measures 8 m.

_____

b. Find the area of a square where each side measures 8 m.

_____

Review these examples for Objective 3:

4.

5 in
13 in
10 in
6 in
3 in
11 in

a. Find the perimeter of the figure above.

Find the perimeter by adding up the lengths of the sides.

Now Try:

4.

5 cm
2 cm
2 cm
3 cm
5 cm
7 cm

a. Find the perimeter of the figure above.

_____

$P = 13$ in. $+ 5$ in. $+ 10$ in. $+ 6$ in. $+ 3$ in. $+ 11$ in.

$P = 48$ in.

b. Find the area of the figure above.

Break up the figure into two pieces. Use just the measurements for the length and width of each piece.
Either draw a line extending the 6 in. side across the figure, or draw a line extending the 10 in side down the figure.

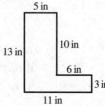

Here we will extend the 6 in. side across the figure, so we have two rectangles: one that is 5 in. by 10 in, and one that is 11 in. by 3 in.

$A = l \cdot w$ $A = l \cdot w$

$A = 5$ in. $\cdot 10$ in. $A = 11$ in. $\cdot 3$ in.

$A = 50$ in.^{2} $A = 33$ in.^{2}

Total area $= 50$ in.$^2 + 33$ in.$^2 = 83$ in.^{2}

b. Find the area of the figure above.

_____

Objective 1   Find the perimeter and area of a rectangle.

For extra help, see Examples 1–2 on pages 546–548 of your text and Section Lecture video for Section 8.3 and Exercise Solutions Clip 3.

Find the perimeter and area of each rectangle.

1.   17 inches by 12 inches   1. _____

2. $4\frac{1}{2}$ yards by $6\frac{1}{2}$ yards   2. _____

3.   37.4 centimeters by 103.2 centimeters   3. _____

Name: Date:

Instructor: Section:

Objective 2   Find the perimeter and area of a square.

For extra help, see Example 3 on pages 549–550 of your text and Section Lecture video for Section 8.3 and Exercise Solutions Clip 11.

Find the perimeter and area of each square with the given side.

4.   7.8 feet   4. _____

5.   8.2 km   5. _____

6.   7.4 inches   6. _____

Objective 3   Find the perimeter and area of a composite figure.

For extra help, see Example 4 on pages 550–551 of your text and Section Lecture video for Section 8.3 and Exercise Solutions Clip 13 and 15.

Find the perimeter and area of each figure. All angles that appear to be right angles are, in fact, right angles.

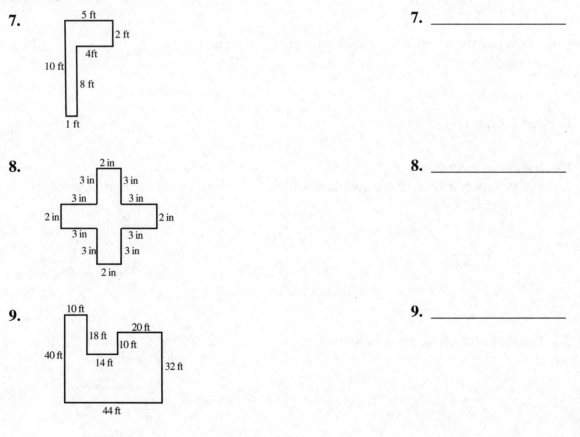

7.   7. _____

8.   8. _____

9.   9. _____

Chapter 8 GEOMETRY

8.4   Parallelograms and Trapezoids

Learning Objectives
1   Find the perimeter and area of a parallelogram.
2   Find the perimeter and area of a trapezoid.

Key Terms

Use the vocabulary terms listed below to complete each statement in exercises 1–4.

    perimeter   area   parallelogram   trapezoid

1.   A _____ is a four-sided figure with both pairs of opposite sides parallel and equal in length.

2.   A _____ is a four-sided figure with exactly one pair of parallel sides.

3.   The formula $P = 2 \cdot l + 2 \cdot w$ is the formula for the _____ of a rectangle.

4.   Square the length of a side of a square to find the _____ of a square.

Guided Examples

Review these examples for Objective 1:

1.   Find the perimeter of the parallelogram.

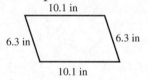

$P = 6.3 \text{ in.} + 10.1 \text{ in.} + 6.3 \text{ in.} + 10.1 \text{ in.}$

$P = 32.8 \text{ in.}$

2.   Find the area of each parallelogram.

a.

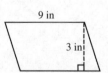

The base is 9 in. and the height is 3 in. Use the formula $A = b \cdot h$.

Now Try:

1.   Find the perimeter of the parallelogram.

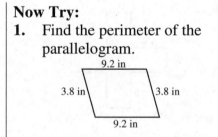

_____

2.   Find the area of each parallelogram.

a.

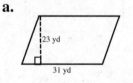

_____

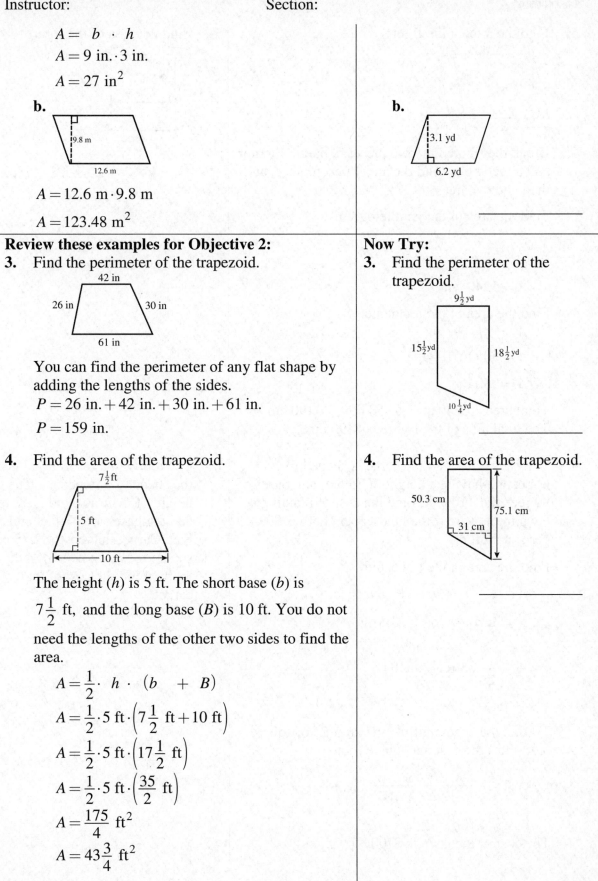

$A = b \cdot h$

$A = 9 \text{ in.} \cdot 3 \text{ in.}$

$A = 27 \text{ in}^2$

b.

$A = 12.6 \text{ m} \cdot 9.8 \text{ m}$

$A = 123.48 \text{ m}^2$

b.

3.1 yd

6.2 yd

_____

Review these examples for Objective 2:

3. Find the perimeter of the trapezoid.

42 in

26 in   30 in

61 in

You can find the perimeter of any flat shape by adding the lengths of the sides.

$P = 26 \text{ in.} + 42 \text{ in.} + 30 \text{ in.} + 61 \text{ in.}$

$P = 159 \text{ in.}$

4. Find the area of the trapezoid.

$7\frac{1}{2}$ ft

5 ft

10 ft

The height (h) is 5 ft. The short base (b) is

$7\frac{1}{2}$ ft, and the long base (B) is 10 ft. You do not

need the lengths of the other two sides to find the area.

$A = \frac{1}{2} \cdot h \cdot (b + B)$

$A = \frac{1}{2} \cdot 5 \text{ ft} \cdot \left(7\frac{1}{2} \text{ ft} + 10 \text{ ft}\right)$

$A = \frac{1}{2} \cdot 5 \text{ ft} \cdot \left(17\frac{1}{2} \text{ ft}\right)$

$A = \frac{1}{2} \cdot 5 \text{ ft} \cdot \left(\frac{35}{2} \text{ ft}\right)$

$A = \frac{175}{4} \text{ ft}^2$

$A = 43\frac{3}{4} \text{ ft}^2$

Now Try:

3. Find the perimeter of the trapezoid.

$9\frac{1}{2}$ yd

$15\frac{1}{2}$ yd $18\frac{1}{2}$ yd

$10\frac{1}{4}$ yd

_____

4. Find the area of the trapezoid.

50.3 cm

75.1 cm

31 cm

_____

5. Find the area of the figure.

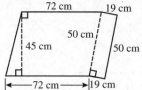

Break the figure into two pieces, a parallelogram and a rectangle. Find the area of each piece, and then add the areas.

Find the area of the parallelogram.

$A = b \cdot h$

$A = 72 \text{ cm} \cdot 45 \text{ cm}$

$A = 3240 \text{ cm}^2$

Find the area of the rectangle.

$A = l \cdot w$

$A = 50 \text{ cm} \cdot 19 \text{ cm}$

$A = 950 \text{ cm}^2$

Total area = 3240 cm^2 + 950 cm^2 = 4190 cm^2
The total area of the figure is 4190 cm^2.

6. The backyard of a new home is shaped like a trapezoid, having a height of 35 feet and bases of 90 feet and 110 feet. Find the cost of planting a lawn in the yard if the landscaper charges $0.20 per square foot.

Find the area of the trapezoid.

$A = \dfrac{1}{2} \cdot h \cdot (b + B)$

$A = \dfrac{1}{2} \cdot 35 \text{ ft} \cdot (90 \text{ ft} + 110 \text{ ft})$

$A = \dfrac{1}{\cancel{2}} \cdot 35 \text{ ft} \cdot \left(\overset{100}{\cancel{200}} \text{ ft}\right)$

$A = 3500 \text{ ft}^2$

To find the cost, multiply the number of square feet times the cost per square foot.

$\text{cost} = \dfrac{3500 \ \cancel{\text{ft}^2}}{1} \cdot \dfrac{\$0.20}{1 \ \cancel{\text{ft}^2}}$

$\text{cost} = \$700$
The cost of the lawn is $700.

5. Find the area of the figure.

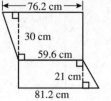

_____

6. The lobby in a resort hotel is in the shape of a trapezoid. The height of the trapezoid is 52 feet and the bases are 47 feet and 59 feet. Carpet that costs $2.75 per square foot is to be laid in the lobby. Find the cost of the carpet.

_____

Name: _____   Date: _____

Instructor: _____   Section: _____

Objective 1   Find the perimeter and area of a parallelogram.

For extra help, see Examples 1–2 on pages 556–557 of your text and Section Lecture video for Section 8.4 and Exercise Solutions Clip 5 and 11.

Find the perimeter of the parallelogram.

1.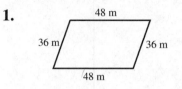

1. _____

Find the area of each parallelogram.

2.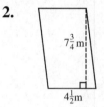

2. _____

Solve the application problem.

3.   A parallelogram has a height of $15\frac{1}{2}$ feet and a base of 20 feet. Find the area.

3. _____

Objective 2   Find the perimeter and area of a trapezoid.

For extra help, see Examples 3–6 on pages 557–559 of your text and Section Lecture video for Section 8.4 and Exercise Solutions Clip 7 and 13.

Find the perimeter of the figure.

4.
```
        276.2 cm
78.6 cm /‾‾‾‾‾‾‾‾\
       /          \ 61 cm
      /_____\
          293 cm
```

4. _____

Find the area of the figure.

5.

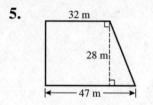

32 m

28 m

47 m

5. _____

Solve the application problem.

6. A swimming pool is in the shape of a parallelogram
with a height of 9.6 meters and base of 12 meters.
Find the cost of a solar pool cover that sells for
$5.10 per square meter.

6. _____

Chapter 8 GEOMETRY

8.5   Triangles

Learning Objectives
1   Find the perimeter of a triangle.
2   Find the area of a triangle.
3   Given the measures of two angles in a triangle, find the measure of the third angle.

Key Terms

Use the vocabulary terms listed below, along with the figure, to complete each statement in exercises 1–3.

base   **height**   **triangle**

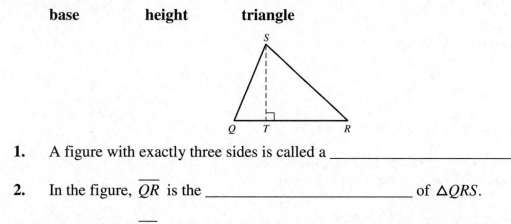

1.   A figure with exactly three sides is called a _____.

2.   In the figure, $\overline{QR}$ is the _____ of $\triangle QRS$.

3.   In the figure, $\overline{ST}$ is the _____ of $\triangle QRS$.

Guided Examples

Review this example for Objective 1:
1.   Find the perimeter of the triangle.

To find the perimeter of a triangle, add the lengths of the three sides.

   $P = 8 \text{ yd} + 6 \text{ yd} + 11 \text{ yd}$

   $P = 25 \text{ yd}$

Now Try:
1.   Find the perimeter of the triangle.

25.7 cm   13.7 cm
      19.6 cm

   _____

Review these examples for Objective 2:

2.   Find the area of each triangle.

a.

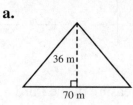

The base is 70 m and the height is 36 m.

$$A = \frac{1}{2} \cdot b \cdot h$$

$$A = \frac{1}{\overset{\displaystyle 2}{2}} \cdot \overset{35}{\cancel{70}} \text{ m} \cdot 36 \text{ m}$$

$$A = 1260 \text{ m}^2$$

b.

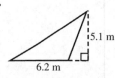

The base must be extended to draw the height.
However, still use 7 yd for b in the formula.
Because the measurements are decimal numbers,
it is easier to use 0.5 in the formula.

$$A = 0.5 \cdot 6.2 \text{ m} \cdot 5.1 \text{ m}$$

$$A = 15.81 \text{ m}^2$$

c.

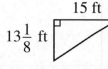

Two sides of the triangle are perpendicular to
each other, so use those sides as the base and the
height.

$$A = \frac{1}{2} \cdot 15 \text{ ft} \cdot 13\frac{1}{8} \text{ ft}$$

$$A = \frac{1}{2} \cdot \frac{15 \text{ ft}}{1} \cdot \frac{105 \text{ ft}}{8}$$

$$A = 98\frac{7}{16} \text{ ft}^2$$

Now Try:

2.   Find the area of each triangle.

a.

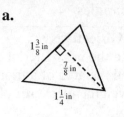

_____

b.

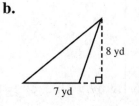

_____

c.

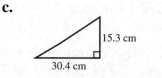

_____

3. Find the area of the shaded part of this figure.

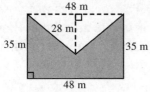

The entire figure is a rectangle. Find the area of the rectangle.

$$A = l \cdot w$$
$$A = 48 \text{ m} \cdot 35 \text{ m}$$
$$A = 1680 \text{ m}^2$$

The unshaded part is a triangle. Find the area of the triangle.

$$A = \frac{1}{2} \cdot b \cdot h$$

$$A = \frac{1}{\cancel{2}_1} \cdot \cancel{48}^{24} \text{ m} \cdot 28 \text{ m}$$

$$A = 672 \text{ m}^2$$

Subtract to find the area of the shaded part.

$$A = \overbrace{1680 \text{ m}^2}^{\text{Entire area}} - \overbrace{672 \text{ m}^2}^{\text{Unshaded part}} = \overbrace{1008 \text{ m}^2}^{\text{Shaded part}}$$

The area of the shaded part is 1008 m^2.

4. For a quilt, Cynthia cuts a triangular section out of a rectangular piece of fabric using the measurements shown above in Example 3, but in centimeters. If the fabric costs \$0.01 per square centimeter, how much does the fabric cost for the section? What is the cost of the fabric that is not used?

From Example 3 above, and changing the units to centimeters, the area of the triangle is 672 cm^2. Multiply the area of the triangular section times the cost per square centimeter.

$$\text{cost} = \frac{672 \text{ cm}^2}{1} \cdot \frac{\$0.01}{1 \text{ cm}^2} = \$6.72$$

The fabric that is not used is the shaded part from Example 3. The unused area, changing the units to centimeters, is 1008 cm^2.

$$\text{cost} = \frac{1008 \text{ cm}^2}{1} \cdot \frac{\$0.01}{1 \text{ cm}^2} = \$10.08$$

The cost of the triangular section is \$6.72. The unused fabric costs \$10.08.

3. Find the area of the shaded part of this figure.

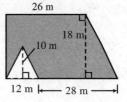

_____

4. For a stained glass window, Bill cuts a triangular section out of the trapezoidal piece of glass using the measurements shown above in Example 3, except in centimeters, rather than meters. If the glass costs \$0.08 per square centimeter, how much does the glass cost for the section? What is the cost of the glass that is not used?

_____

Review these examples for Objective 3:

5. The measures of two angles of a triangle are given. Find the measure of the third angle.

   a. 49°, 72°

   Step 1 Add the two angle measurements you are given.
   $$49° + 72° = 121°$$

   Step 2 Subtract the sum from 180°.
   $$180° - 121° = 59°$$

   The measure of the third angle is 59°.
   b. 50°, 90°

   The second angle is a right angle since it measures 90°.

   Step 1 $50° + 90° = 140°$
   Step 2 $180° - 140° = 40°$

   The measure of the third angle is 40°.

Now Try:

5. The measures of two angles of a triangle are given. Find the measure of the third angle.
   a. 87°, 13°

   _____

   b. 90°, 25°

   _____

Objective 1   Find the perimeter of a triangle.

For extra help, see Example 1 on page 562 of your text and Section Lecture video for Section 8.5 and Exercise Solutions Clip 5a and 7a.

Find the perimeter of each triangle.

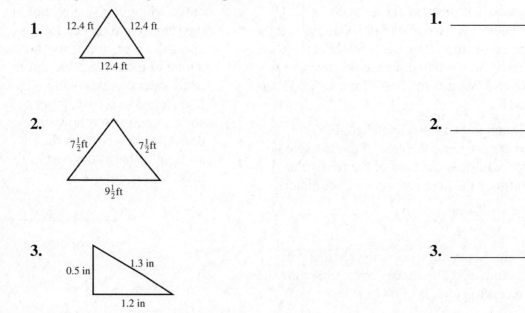

1. 12.4 ft 12.4 ft 12.4 ft

2. $7\frac{1}{2}$ ft $7\frac{1}{2}$ ft $9\frac{1}{2}$ ft

3. 0.5 in 1.3 in 1.2 in

1. _____

2. _____

3. _____

Name: Date:
Instructor: Section:

Objective 2   Find the area of a triangle.

For extra help, see Examples 2–4 on pages 563–564 of your text and Section Lecture video for Section 8.4 and Exercise Solutions Clip 5b and 7b.

Find the area of each triangle.

4.

9 ft

6 ft

4. _____

5.

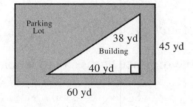

16 cm

14 cm

5. _____

Find the shaded area in the figure.

6.

Parking Lot

38 yd
Building
45 yd
40 yd
60 yd

6. _____

Objective 3   Given the measures of two angles in a triangle, find the measure of the third angle.

For extra help, see Example 5 on page 565 of your text and Section Lecture video for Section 8.5 and Exercise Solutions Clip 17.

The measures of two angles of a triangle are given. Find the measure of the third angle.

7.   37°, 62°

7. _____

8.   51°, 78°

8. _____

9.   76°, 76°

9. _____

Chapter 8 GEOMETRY

8.6   Circles

Learning Objectives	
1	Find the radius and diameter of a circle.
2	Find the circumference of a circle.
3	Find the area of a circle.
4	Become familiar with Latin and Greek prefixes used in math terminology.

Key Terms

Use the vocabulary terms listed below to complete each statement in exercises 1–5.

       circle   **radius**   **diameter**   **circumference**   **π (pi)**

1.   The _____ is the distance from the center of a circle to any point on the circle.

2.   The _____ of a circle is the distance around the circle.

3.   A figure with all points the same distance from a fixed center point is called a

     _____.

4.   The ratio of the circumference to the diameter of any circle equals _____.

5.   The _____ of a circle is the distance across a circle, passing through the center.

Guided Examples

Review these examples for Objective 1:

1.   Find the unknown length of the diameter or radius of each circle.

     a.

Because the radius is 64 ft, the diameter is twice as long.

$$d = 2 \cdot r$$

$$d = 2 \cdot 64 \text{ ft}$$

$$d = 128 \text{ ft}$$

Now Try:

1.   Find the unknown length of the diameter or radius of each circle.

     a.

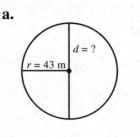

                _____

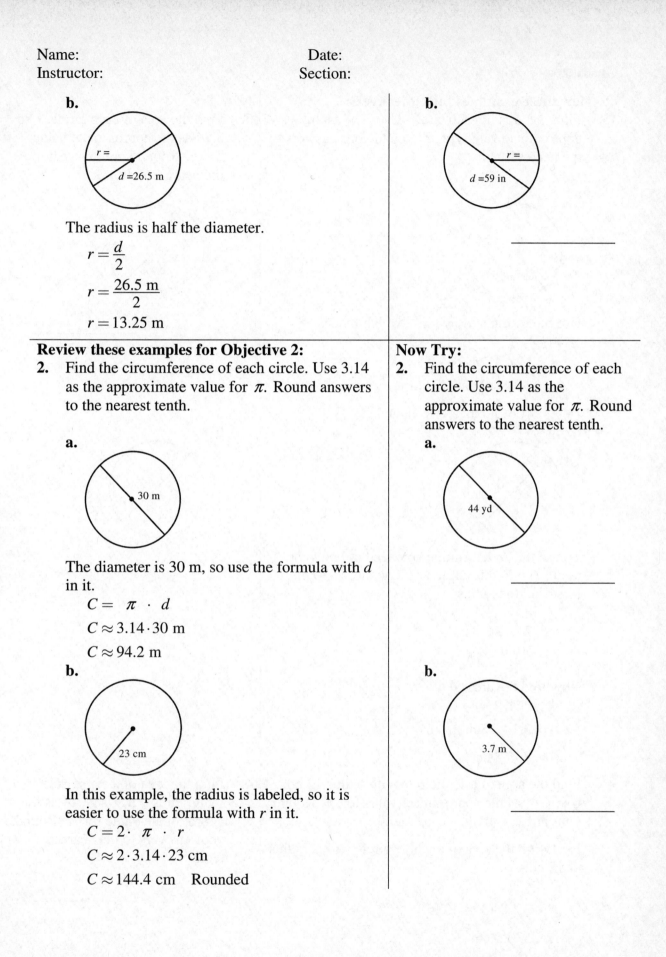

b.

The radius is half the diameter.

$$r = \frac{d}{2}$$

$$r = \frac{26.5 \text{ m}}{2}$$

$$r = 13.25 \text{ m}$$

b.

_____

Review these examples for Objective 2:

2. Find the circumference of each circle. Use 3.14 as the approximate value for π. Round answers to the nearest tenth.

a.

The diameter is 30 m, so use the formula with d in it.

$$C = \pi \cdot d$$

$$C \approx 3.14 \cdot 30 \text{ m}$$

$$C \approx 94.2 \text{ m}$$

b.

In this example, the radius is labeled, so it is easier to use the formula with r in it.

$$C = 2 \cdot \pi \cdot r$$

$$C \approx 2 \cdot 3.14 \cdot 23 \text{ cm}$$

$$C \approx 144.4 \text{ cm} \quad \text{Rounded}$$

Now Try:

2. Find the circumference of each circle. Use 3.14 as the approximate value for π. Round answers to the nearest tenth.

a.

_____

b.

_____

Review these examples for Objective 3:

3. Find the area of each circle. Use 3.14 as the approximate value for π. Round your answers to the nearest tenth.

a.

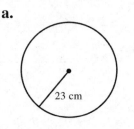

Use the formula $A = \pi \cdot r^2$, which means $\pi \cdot r \cdot r$.

$$A = \pi \cdot r \cdot r$$

$$A \approx 3.14 \cdot 23 \text{ cm} \cdot 23 \text{ cm}$$

$$A \approx 1661.1 \text{ cm}^2 \quad \text{Rounded}$$

b.

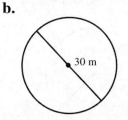

To use the area formula, you need to know the radius (r). In this circle, the diameter is 30 m. First find the radius.

$$r = \frac{d}{2}$$

$$r = \frac{30 \text{ m}}{2} = 15 \text{ m}$$

Now find the area.

$$A = \pi \cdot r \cdot r$$

$$A \approx 3.14 \cdot 15 \text{ m} \cdot 15 \text{ m}$$

$$A \approx 706.5 \text{ m}^2$$

4. Find the area of a semicircle with radius 13 cm. Use 3.14 as the approximate value for π. Round your answers to the nearest tenth.

First, find the area of a whole circle with a radius of 13 cm.

$$A = \pi \cdot r \cdot r$$

$$A \approx 3.14 \cdot 13 \text{ m} \cdot 13 \text{ m}$$

$$A \approx 530.66 \text{ m}^2$$

Now Try:

3. Find the area of each circle. Use 3.14 as the approximate value for π. Round your answers to the nearest tenth.

a.

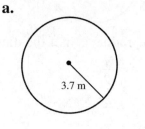

_____

b.

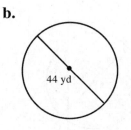

_____

4. Find the area of a semicircle with radius 9 ft. Use 3.14 as the approximate value for π. Round your answers to the nearest tenth.

_____

Divide the area of the whole circle by 2 to find the area of the semicircle.

$$\frac{530.66 \text{ m}^2}{2} = 265.33 \text{ m}^2$$

The last step is rounding 265.33 to the nearest tenth.

Area of semicircle $\approx 265.3 \text{ m}^2$ Rounded

5. A circular piece of glass, for a customized table, is 3 ft in diameter. The cost of metal for the edge is $3.50 per foot. What will it cost to add a metal edge to the glass? Use 3.14 for π.

Circumference $= \pi \cdot d$

$C \approx 3.14 \cdot 3 \text{ ft}$

$C \approx 9.42 \text{ ft}$

cost = cost per foot · circumference

$\text{cost} = \frac{\$3.50}{1 \text{ ft}} \cdot \frac{9.42 \text{ ft}}{1}$

cost = $32.97

The cost of adding a metal edge to the glass is $32.97.

6. Find the cost of covering the glass in Example 5 with a plastic cover. The material for the cover costs $2.50 per square foot. Use 3.14 for π. Round your answer to the nearest cent.

First find the radius. $r = \frac{d}{2} = \frac{3 \text{ ft}}{2} = 1.5 \text{ ft}$

Then find the area. $A = \pi \cdot r^2$

$A \approx 3.14 \cdot 1.5 \text{ ft} \cdot 1.5 \text{ ft}$

$A \approx 7.065 \text{ ft}^2$

$\text{cost} = \frac{\$2.50}{1 \text{ ft}^2} \cdot \frac{7.065 \text{ ft}^2}{1}$

cost = $17.66   Rounded

The cost of the plastic cover is $17.66.

Review this example for Objective 4:

7. Write one math term and one nonmathematical term that use the prefix "*tri-*" (three).

tri- (three): triangle; tricycle

Other answers are possible.

5. A circular coaster is 5 inches in diameter. The cost of rubber for the edge is $0.80 per inch. What will it cost to add rubber edge to the coaster? Use 3.14 for π.

_____

6. Find the cost of covering the coaster in Example 5 with a protective coating. The material for the cover costs $0.10 per square inch. Use 3.14 for π. Round your answer to the nearest cent.

_____

Now Try:

7. Write one math term and one nonmathematical term that use the prefix "*milli-*" (thousand).

_____

Objective 1   Find the radius and diameter of a circle.

For extra help, see Example 1 on pages 569–570 of your text and Section Lecture video for Section 8.6 and Exercise Solutions Clip 3 and 5.

Find the diameter or radius in each circle.

1.   The diameter of a circle is 8 feet. Find its radius.   1. _____

2.   The radius of a circle is 2.7 centimeters. Find its   2. _____
     diameter.

3.   The diameter of a circle is $12\frac{1}{2}$ yards. Find its   3. _____
     radius.

Objective 2   Find the circumference of a circle.

For extra help, see Example 2 on page 571 of your text and Section Lecture video for Section 8.6 and Exercise Solutions Clip 7 and 9.

Find the circumference of each circle. Use 3.14 as an approximation for π. Round each answer to the nearest tenth.

4.   A circle with a diameter of $4\frac{3}{4}$ inches   4. _____

5.   A circle with a radius of 4.5 yards   5. _____

6.   A circle with a radius of 8 cm   6. _____

Objective 3   Find the area of a circle.

For extra help, see Examples 3–6 on page 572–574 of your text and Section Lecture video for Section 8.6 and Exercise Solutions Clip 7 and 9.

Find the area of each circle. Use 3.14 as an approximation for π. Round each answer to the nearest tenth.

7.   A circle with diameter of $5\frac{1}{3}$ yards

7. _____

8.   A circle with radius of 9.8 centimeters

8. _____

9.

9. _____

Objective 4   Become familiar with Latin and Greek prefixes used in math terminology.

For extra help, see Example 7 on page 574 of your text and Section Lecture video for Section 8.6 and Exercise Solutions Clip 11.

Write one math term and one nonmathematical term that use each prefix listed below. (Answers will vary.)

10.   *oct*-(eight)

10. _____

11.   *poly*-(many)

11. _____

12.   *uni*-(one)

12. _____

Chapter 8 GEOMETRY

8.7   Volume

Learning Objectives
Find the volume of a
1   rectangular solid;
2   sphere;
3   cylinder;
4   cone and pyramid.

Key Terms

Use the vocabulary terms listed below to complete each statement in exercises 1–6.

volume   rectangular solid   sphere

cylinder   cone   pyramid

1.   A _____ is a box-like solid figure.

2.   A solid figure with two congruent, parallel, circular bases is a _____.

3.   A _____ is a ball-like solid figure.

4.   _____ is a measure of the space inside a solid shape.

5.   A solid figure whose base is a square or a rectangle and whose faces (sides) are triangles is called a _____.

6.   A solid figure with only one base, and that base is a circle, is called a _____.

Guided Examples

Review this example for Objective 1:
1.   Find the volume of the box.

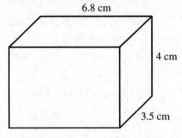

Use the formula $V = l \cdot w \cdot h$.
$V = 6.8 \text{ cm} \cdot 3.5 \text{ cm} \cdot 4 \text{ cm}$

$V = 95.2 \text{ cm}^3$

Now Try:
1.   Find the volume of the box.

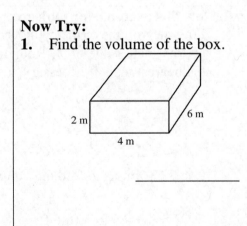

_____

Name: Date:
Instructor: Section:

Review these example for Objective 2:

2. Find the volume of the sphere with the help of a
    calculator. Use 3.14 as the approximate value of
    π. Round your answer to the nearest tenth.

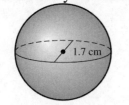

The diameter is 1.7 cm, so the radius is
$\frac{1.7\text{cm}}{2} = 0.85\text{cm}$.

Find the volume.

$$V = \frac{4}{3} \cdot \pi \cdot r^3$$

$$V \approx \frac{4 \cdot 3.14 \cdot 0.85 \text{ cm} \cdot 0.85 \text{ cm} \cdot 0.85 \text{ cm}}{3}$$

$$V \approx 2.571136667 \text{ cm}^3$$

$$V \approx 2.6 \text{ cm}^3$$

3. Find the volume of the hemisphere with the help
    of a calculator. Use 3.14 for π. Round your
    answer to the nearest tenth.

$$V = \frac{2 \cdot \pi \cdot r^3}{3}$$

$$V \approx \frac{2 \cdot 3.14 \cdot 3 \text{ ft} \cdot 3 \text{ ft} \cdot 3 \text{ ft}}{3}$$

$$V \approx 56.5 \text{ ft}^3 \quad \text{Rounded to the nearest tenth.}$$

Review this example for Objective 3:

4. Find the volume of the cylinder. Use 3.14 as the
    approximate value of π. Round your answer to
    the nearest tenth if necessary.

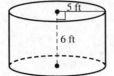

Use the formula to find the volume.

$$V = \pi \cdot r^2 \cdot h$$

$$V \approx 3.14 \cdot 5 \text{ ft} \cdot 5 \text{ ft} \cdot 6 \text{ ft}$$

$$V \approx 471 \text{ ft}^3$$

Now Try:

2. Find the volume of the sphere
    with the help of a calculator.
    Use 3.14 as the approximate
    value of π. Round your answer
    to the nearest tenth.

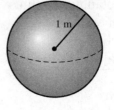

_____

3. Find the volume of the
    hemisphere with the help of a
    calculator. Use 3.14 for π.
    Round your answer to the
    nearest tenth.

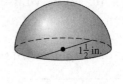

_____

Now Try:

4. Find the volume of the cylinder.
    Use 3.14 as the approximate
    value of π. Round your answer
    to the nearest tenth if necessary.

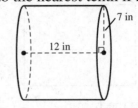

_____

Review these examples for Objective 4:

5. Find the volume of the cone. Use 3.14 for π. Round your answer to the nearest tenth.

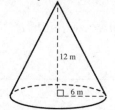

First find the value of B in the formula, which is the area of the circular base. Recall that the formula for the area of the circle is πr^2.

$$B = \pi \cdot r \cdot r$$
$$B \approx 3.14 \cdot 6\ m \cdot 6\ m$$
$$B \approx 113.04\ m^2 \leftarrow \text{Do not round to tenths yet.}$$

Now find the volume. The height is 12 m.

$$V = \frac{B \cdot h}{3}$$
$$V \approx \frac{113.04\ m^2 \cdot 12\ m}{3}$$
$$V \approx 452.16\ m^3$$
$$V \approx 452.2\ m^3$$

6. Find the volume of this pyramid with rectangular base. Round your answer to the nearest tenth.

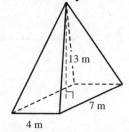

First find the value of B in the formula, which is the area of a rectangular base. Recall that the area of the rectangle is found by multiplying length times width.

$$B = 7\ m \cdot 4\ m$$
$$B = 28\ m^2$$

Now find the volume.

$$V = \frac{B \cdot h}{3}$$
$$V \approx \frac{28\ m^2 \cdot 13\ m}{3}$$
$$V \approx 121.3\ m^3$$

Now Try:

5. Find the volume of the cone. Use 3.14 for π. Round your answer to the nearest tenth.

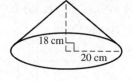

_____

6. Find the volume of this pyramid with rectangular base. Round your answer to the nearest tenth.

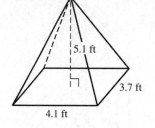

_____

Name: Date:

Instructor: Section:

Objective 1   Find the volume of a rectangular solid.

For extra help, see Example 1 on page 582 of your text and Section Lecture video for Section 8.7 and Exercise Solutions Clip 3.

Find the volume of each rectangular solid. Round answers to the nearest tenth, if necessary.

1.

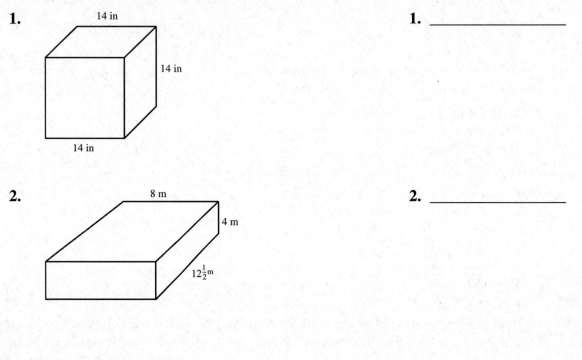

1. _____

2.

2. _____

3.

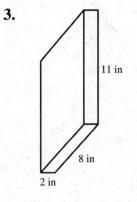

3. _____

Objective 2   Find the volume of a sphere.

For extra help, see Examples 2–3 on page 583–584 of your text and Section Lecture video for Section 8.7 and Exercise Solutions Clip 7.

Find the volume of each sphere or hemisphere. Use 3.14 as an approximation for π. Round answers to the nearest tenth, if necessary.

4.   A sphere with a diameter of $3\frac{1}{4}$ inches.   4. _____

5.   A hemisphere with a radius of 11.6 feet.   5. _____

6.   A sphere with a radius of 6.8 cm   6. _____

Objective 3   Find the volume of a cylinder.

For extra help, see Examples 4 on page 584 of your text and Section Lecture video for Section 8.7 and Exercise Solutions Clip 9.

Find the volume of each figure. Use 3.14 as an approximation for π. Round answers to the nearest tenth, if necessary.

7.   0.2 km   7. _____

       4 km

8.   A coffee can, radius 6 centimeters and height 16   8. _____
      centimeters

9.   A cardboard mailing tube, diameter 5 centimeters   9. _____
      and height 25 centimeters

Name: Date:
Instructor: Section:

Objective 4   Find the volume of a cone and a pyramid.

For extra help, see Examples 5–6 on page 585–586 of your text and Section Lecture video for Section 8.7 and Exercise Solutions Clip 13.

Find the volume of each figure. Use 3.14 as an approximation for π. Round answers to the nearest tenth, if necessary.

10.   10. _____

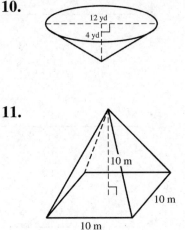

11.   11. _____

12.   Find the volume of a pyramid with square base 42   12. _____
       meters on a side and height 38 meters.

Chapter 8 GEOMETRY

8.8   Pythagorean Theorem

Learning Objectives
1   Find square roots using the square root key on a calculator.
2   Find the unknown length in a right triangle.
3   Solve application problems involving right triangles.

Key Terms

Use the vocabulary terms listed below to complete each statement in exercises 1–3.

hypotenuse   legs   right triangle

1.   A triangle with a 90° angle is called a _____.

2.   The side opposite the right angle in a right triangle is called the
       _____ of the triangle.

3.   The two sides of the right angle in a right triangle are called the
       _____ of the triangle.

Guided Examples

Review these examples for Objective 1:
1.   Use a calculator to find each square root. Round answers to the nearest hundredth.

    a. $\sqrt{45}$

Calculator shows 6.708203932; round to 6.71.

    b. $\sqrt{86}$

Calculator shows 9.273618495; round to 9.27.

    c. $\sqrt{120}$

Calculator shows 10.95445115; round to 10.95.

Now Try:
1.   Use a calculator to find each square root. Round answers to the nearest hundredth.

    a. $\sqrt{20}$

    _____

    b. $\sqrt{92}$

    _____

    c. $\sqrt{134}$

    _____

Review these examples for Objective 2:

2. Find the unknown length in each right triangle.
    Round answers to the nearest tenth if necessary.

a.

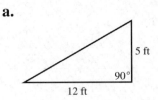

The unknown length is the side opposite the
right angle, which is the hypotenuse. Use the
formula for finding the hypotenuse.

$$\text{hypotenuse} = \sqrt{(\text{leg})^2 + (\text{leg})^2}$$

$$\text{hypotenuse} = \sqrt{(5)^2 + (12)^2}$$

$$= \sqrt{25 + 144}$$

$$= \sqrt{169}$$

$$= 13$$

The hypotenuse is 13 ft long.

b.

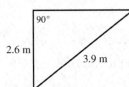

We do know the length of the hypotenuse
(3.9 m), so it is the length of one of the legs that
is unknown. Use the formula for finding the leg.

$$\text{leg} = \sqrt{(\text{hypotenuse})^2 - (\text{leg})^2}$$

$$\text{leg} = \sqrt{(3.9)^2 - (2.6)^2}$$

$$= \sqrt{15.21 - 6.76}$$

$$= \sqrt{8.45}$$

$$\approx 2.9$$

The length of the leg is approximately 2.9 m.

Now Try:

2. Find the unknown length in each
    right triangle. Round answers to
    the nearest tenth if necessary.

a.

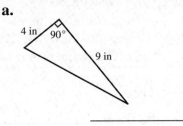

_____

b.

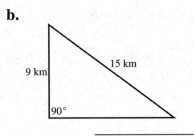

_____

Name: _____   Date: _____
Instructor: _____   Section: _____

Review this example for Objective 3:

3. The base of a ladder is located 7 feet from a building. The ladder reaches 24 feet up the building. How long is the ladder? Round to the nearest tenth of a foot if necessary.

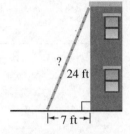

A right triangle is formed. The unknown side is the hypotenuse.

$$\text{hypotenuse} = \sqrt{(\text{leg})^2 + (\text{leg})^2}$$

$$\text{hypotenuse} = \sqrt{(7)^2 + (24)^2}$$

$$= \sqrt{49 + 576}$$

$$= \sqrt{625}$$

$$= 25$$

The length of the ladder is 25 ft.

Now Try:

3. Find the unknown length in this roof plan. Round to the nearest tenth of a foot if necessary.

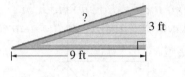

_____

Objective 1   Find square roots using the square root key on a calculator.

For extra help, see Example 1 on page 590 of your text and Section Lecture video for Section 8.8.

Find each square root. Use a calculator with a square root key. Round the answer to the nearest thousandth, if necessary.

1. $\sqrt{17}$

2. $\sqrt{75}$

3. $\sqrt{102}$

1. _____

2. _____

3. _____

Name: Date:

Instructor: Section:

Objective 2   Find the unknown length in a right triangle.

For extra help, see Example 2 on page 591 of your text and Section Lecture video for Section 8.8 and Exercise Solutions Clip 19 and 23.

Find the unknown length in each right triangle. Use a calculator with a square root key. Round the answer to the nearest tenth, if necessary.

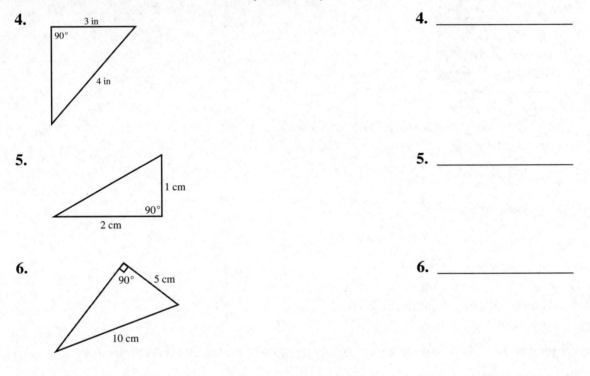

4.

4. _____

5.

5. _____

6.

6. _____

Objective 3   Solve application problems involving right triangles.

For extra help, see Example 3 on page 592 of your text and Section Lecture video for Section 8.8 and Exercise Solutions Clip 43 and 45.

Solve each application problem. Draw a diagram if one is not provided. Use a calculator with a square root key. Round the answer to the nearest tenth, if necessary.

7.   Find the length of the loading ramp.   **7.** _____

8.   A kite is flying on 50 feet of string. If the horizontal   **8.** _____
distance of the kite from the person flying it is 40
feet, how far off the ground is the kite?

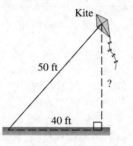

9.   The base of a 17-ft ladder is located 15 ft from a   **9.** _____
building. How high up on the building will the
ladder reach?

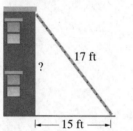

Chapter 8 GEOMETRY

8.9   Similar Triangles

Learning Objectives
1   Identify corresponding parts in similar triangles.
2   Find the unknown lengths of sides in similar triangles.
3   Solve application problems involving similar triangles.

Key Terms

Use the vocabulary terms listed below to complete each statement in exercises 1–2.

similar triangles   congruent

1.   Two angles are _____ if their measures are equal.

2.   _____ are triangles with the same shape but not necessarily the same size.

Guided Examples

Review this example for Objective 1:
1.   Identify corresponding angles and sides in these similar triangles.

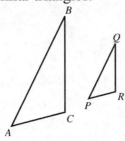

$\overline{AB}$ and $\overline{PQ}$; $\overline{AC}$ and $\overline{PR}$; $\overline{BC}$ and $\overline{QR}$;

$\angle A$ and $\angle P$; $\angle B$ and $\angle Q$; $\angle C$ and $\angle R$

Now Try:
1.   Identify corresponding angles and sides in these similar triangles.

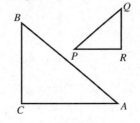

_____

Review these examples for Objective 2:

2. Find the length of b in the smaller triangle. Assume the triangles are similar.

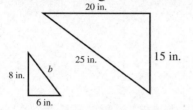

The length you want to find in the smaller triangle is side b, and it corresponds to 25 in. in the larger triangle. The smaller triangle is turned "upside down" compared to the larger triangle, so be careful when identifying corresponding sides. Then notice that 6 in. in the smaller triangle corresponds to 15 in. in the larger triangle, and you know both of their lengths. Because the ratios of the lengths of corresponding sides are equal, you can set up a proportion.

$$\frac{b}{25} = \frac{6}{15}$$

$$\frac{b}{25} = \frac{2}{5} \quad \text{Write } \frac{6}{15} \text{ in lowest terms as } \frac{2}{5}.$$

Find the cross products.

$$b \cdot 5 = 25 \cdot 2$$

$$\frac{b \cdot \cancel{5}^{1}}{\cancel{5}_{1}} = \frac{50}{5}$$

$$b = 10$$

Side b has length of 10 in.

3. Find the perimeter of the smaller triangle. Assume the triangles are similar.

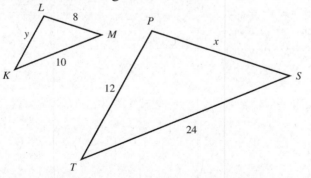

First, find y, the length of $\overline{KL}$, then add the lengths of all three sides to find the perimeter.

Now Try:

2. Find the length of x in the smaller triangle. Assume the triangles are similar.

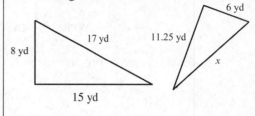

_____

3. Find the perimeter of the smaller triangle. Assume the triangles are similar.

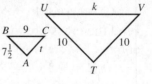

_____

The ratios of the lengths of corresponding sides are equal, so you can set up a proportion.

$$\begin{array}{l} KL \to \\ KM \to \end{array} \dfrac{y}{10} = \dfrac{12}{24} \begin{array}{l} \leftarrow PT \\ \leftarrow TS \end{array}$$

$$\dfrac{y}{10} = \dfrac{1}{2} \quad \text{Write } \dfrac{12}{24} \text{ in lowest terms as } \dfrac{1}{2}.$$

Find the cross products.

$$\dfrac{y}{10} = \dfrac{1}{2}$$

$$y \cdot 2 = 10 \cdot 1$$

$$\dfrac{y \cdot \overset{1}{\cancel{2}}}{\underset{1}{\cancel{2}}} = \dfrac{10}{2}$$

$$y = 5$$

$\overline{KL}$ has a length of 5.

Now add the lengths of all three sides to find the perimeter of the smaller triangle.

$$\text{Perimeter} = 8 + 10 + 5 = 23$$

Review this example for Objective 3:

4. The height of the house shown here can be found by using similar triangles and proportion. Find the height of the house by writing a proportion and solving it.

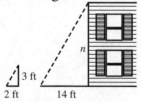

The triangles shown are similar, so write a proportion to find n.

$$\begin{array}{l} \text{height of larger triangle} \to \\ \text{height of smaller triangle} \to \end{array} \dfrac{n}{3} = \dfrac{14}{2} \begin{array}{l} \leftarrow \text{length of larger triangle} \\ \leftarrow \text{length of smaller triangle} \end{array}$$

Find the cross products and show that they are equal.

$$n \cdot 2 = 3 \cdot 14$$

$$n \cdot 2 = 42$$

$$\dfrac{n \cdot \overset{1}{\cancel{2}}}{\underset{1}{\cancel{2}}} = \dfrac{42}{2}$$

$$n = 21$$

The height of the house is 21 ft.

Now Try:

4. A flagpole casts a shadow 77 feet long at the same time that a pole 15 feet tall casts a shadow 55 ft long. Find the height of the flagpole.

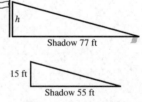

_____

Name: Date:

Instructor: Section:

Objective 1   Identify the corresponding parts in similar triangles.

For extra help, see page 597 of your text and Section Lecture video for Section 8.9 and Exercise Solutions Clip 7 and 9.

Name the corresponding angles and the corresponding sides in each pair of similar triangles.

1.

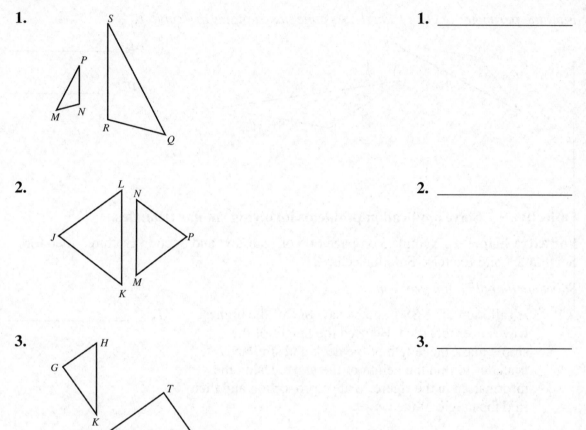

1. _____

2.

2. _____

3.

3. _____

Objective 2   Find the unknown lengths of sides in similar triangles.

For extra help, see Examples 1–2 on pages 598–599 of your text and Section Lecture video for Section 8.9 and Exercise Solutions Clip 11 and 15.

Find the unknown lengths in each pair of similar triangles.

4.

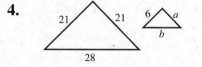

4. _____

5.

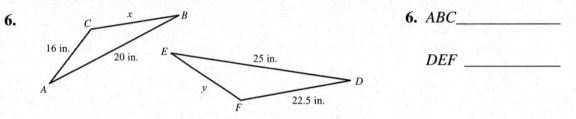

5. _____

Find the perimeter of each triangle. Assume the triangles are similar.

6.

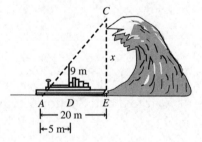

6. *ABC*_____

    DEF _____

Objective 3   Solve application problems involving similar triangles.

For extra help, see Example 3 on page 600 of your text and Section Lecture video for Section 8.9 and Exercise Solutions Clip 21.

Solve each application problem.

7.   A sailor on the USS Ramapo saw one of the highest waves ever recorded. He used the height of the ship's mast, the length of the deck and similar triangles to find the height of the wave. Using the information in the figure, write a proportion and then find the height of the wave.

7. _____

8. A fire lookout tower provides an excellent view of the surrounding countryside. The height of the tower can be found by lining up the top of the tower with the top of a 3-meter stick. Use similar triangles to find the height of the tower.

8. _____

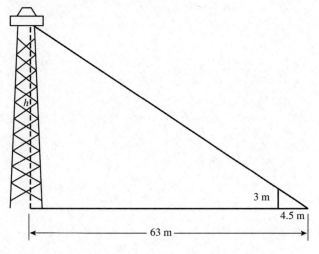

9. A 30 m ladder touches the side of a building at a height of 25 m. At what height would a 12-m ladder touch the building if it makes the same angle with the ground?

9. _____

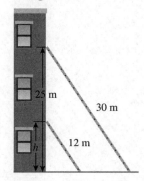

Chapter 9 BASIC ALGEBRA

9.1   Signed Numbers

Learning Objectives
1   Write negative numbers.
2   Graph signed numbers on a number line.
3   Use the < and > symbols.
4   Find absolute value.
5   Find the opposite of a number.

Key Terms

Use the vocabulary terms listed below to complete each statement in exercises 1–4.

negative numbers   signed numbers   absolute value

opposite of a number

1.   _____ include positive numbers, negative numbers, and zero.

2.   The _____ is the same distance from 0 on the number line as the original number, but located on the other side of 0.

3.   The distance from a number to 0 on the number line is called the _____ of the number.

4.   Numbers that are less than 0 are _____.

Guided Examples

Review this example for Objective 2:
1.   **Graph the set of numbers on a number line.**

$$-3, -5, -1\frac{1}{2}, \frac{2}{3}, 0, 4$$

Place a dot at the correct location for each number.

Now Try:
1.   **Graph the set of numbers on a number line.**

$$-5, -4, -3\frac{2}{3}, -1\frac{1}{2}, 0, 2$$

$$\xleftarrow{\quad}\underset{\begin{array}{ccccccccccccccccc}-9&-8&-7&-6&-5&-4&-3&-2&-1&0&1&2&3&4&5&6\end{array}}{\rule{0pt}{0pt}}\xrightarrow{\quad}$$

Review these examples for Objective 3:

2. Use the number line above to compare each pair of numbers. Then write > or < to make each statement true.

a. Compare 3 and 6.

3 < 6 (read "3 is less than 6") because 3 is to the left of 6 on the number line.

b. Compare –7 and –5.

–7 < –5 because –7 is to the left of –5.

c. Compare 4 and –2.

4 > –2 because 4 is to the right of –2.

d. Compare –6 and 0.

–6 < 0 because –6 is to the left of 0.

Now Try:

2. Use the number line above to compare each pair of numbers. Then write > or < to make each statement true.

a. Compare 2 and 5.

_____

b. Compare –8 and –3.

_____

c. Compare 5 and –5.

_____

d. Compare –9 and 0.

_____

Review these examples for Objective 4:

3. Simplify each absolute value expression.

a. $|7|$

The distance from 0 to 7 is 7 spaces, so $|7|=7$.

b. $|-7|$

The distance from 0 to –7 is 7 spaces, so $|-7|=7$.

c. $|0|$

The distance form 0 to 0 is 0 spaces, so $|0|=0$.

d. $-|-4|$

First, $|-4|$ is 4. But there is also a negative sign outside the absolute value bars. So, –4 is the simplified expression.

Now Try:

3. Simplify each absolute value expression.

a. $|6|$

_____

b. $|-6|$

_____

c. $|0|$

_____

d. $-|-8|$

_____

Review these examples for Objective 5:

4. Find the opposite of each number.

   a. 8

The opposite is $-(8) = -8$.

   b. 5

The opposite is $-(5) = -5$.

   c. $\dfrac{3}{4}$

The opposite is $-\left(\dfrac{3}{4}\right) = -\dfrac{3}{4}$.

   d. 0

The opposite is $-(0) = 0$.
The opposite of 0 is 0.

5. Find the opposite of each number.

   a. –3

The opposite is $-(-3) = 3$, by the double negative rule.

   b. –12

The opposite is $-(-12) = 12$.

   c. $-\dfrac{5}{6}$

The opposite is $-\left(-\dfrac{5}{6}\right) = \dfrac{5}{6}$.

Now Try:

4. Find the opposite of each number.

   a. 6

_____

   b. 11

_____

   c. $\dfrac{15}{17}$

_____

   d. 0

_____

5. Find the opposite of each number.

   a. –14

_____

   b. –27

_____

   c. $-\dfrac{7}{8}$

_____

Objective 1   Write negative numbers.

For extra help, see page 626 of your text and Section Lecture video for Section 9.1.

Write a signed number for each situation.

   1. The temperature was 17 degrees above zero.

       1. _____

   2. The lake was 120 feet below sea level.

       2. _____

   3. The football team lost 7 yards on the first play.

       3. _____

Name: Date:

Instructor: Section:

Objective 2   Graph signed numbers on a number line.

For extra help, see Example 1 on page 627 of your text and Section Lecture video for Section 9.1 and Exercise Solutions Clip 9.

Graph each set of numbers on a number line.

4.   –2, –1, 0, 1, 2, 5

4.

$$\xleftarrow{\;+\;+\;+\;+\;+\;+\;+\;+\;+\;+\;}\rightarrow$$
$$-5\;-4\;-3\;-2\;-1\;\;0\;\;1\;\;2\;\;3\;\;4\;\;5$$

5.   –4.5, –1.5, –0.5, 0, 1.5, 2.5

5.

$$\xleftarrow{\;+\;+\;+\;+\;+\;+\;+\;+\;+\;+\;}\rightarrow$$
$$-5\;-4\;-3\;-2\;-1\;\;0\;\;1\;\;2\;\;3\;\;4\;\;5$$

6. $-\dfrac{1}{2}, -3, -\dfrac{5}{2}, \dfrac{1}{4}, 1\dfrac{7}{8}, 3$

6.

$$\xleftarrow{\;+\;+\;+\;+\;+\;+\;+\;+\;+\;+\;}\rightarrow$$
$$-5\;-4\;-3\;-2\;-1\;\;0\;\;1\;\;2\;\;3\;\;4\;\;5$$

Objective 3   Use the < and > symbols.

For extra help, see Example 2 on pages 627–628 of your text and Section Lecture video for Section 9.1 and Exercise Solutions Clip 15, 19, and 23.

Write > or < in each blank to make a true statement.

7.   –6 ____ 0

7. _____

8.   5 ____ –7

8. _____

9.   –5 ____ –3

9. _____

Objective 4   Find absolute value.

For extra help, see Example 3 on page 628 of your text and Section Lecture video for Section 9.1 and Exercise Solutions Clip 27, 29, and 33.

Simplify each absolute value expression.

10. $\left|\dfrac{8}{9}\right|$

10. _____

11. $-\left|-8.23\right|$

11. _____

12. $\left|-11\right|$

12. _____

Objective 5   Find the opposite of a number.

For extra help, see Examples 4–5 on page 629 of your text and Section Lecture video for Section 9.1.

Find the opposite of each number.

13.   3   13. _____

14.   −10   14. _____

15. $-\dfrac{2}{3}$ 15. _____

Chapter 9 BASIC ALGEBRA

9.2   Adding and Subtracting Signed Numbers

Learning Objectives	
1	Add signed numbers by using a number line.
2	Add signed numbers without using a number line.
3	Find the additive inverse of a number.
4	Subtract signed numbers.
5	Add or subtract a series of signed numbers.

Key Terms

Use the vocabulary terms listed below to complete each statement in exercises 1–2.

additive inverse   absolute value

1. $|-4|$ is read as the "_____ of negative four."

2.   The opposite of a number is its _____.

Guided Examples

Review these examples for Objective 1:

1.   Add using a number line.

    a. $5+(-2)$

Start at 0 and draw an arrow 5 units to the right. From the end of this arrow, draw an arrow 2 units to the left. This second arrow ends at 3, so

    $5+(-2)=3$

    b. $-4+3$

Draw an arrow from 0 going 4 units to the left. From the end of this arrow, draw an arrow 3 units to the right. This second arrow ends at –1, so

    $-4+3=-1$

Now Try:

1.   Add using a number line.

    a. $6+(-3)$

    _____

    b. $-8+5$

    _____

c. $-2+(-4)$

As the arrows along the number line show,

$$-2+(-4)=-6$$

c. $-4-5$

_____

Review these examples for Objective 2:

2. Add (without using number lines).

a. $-5+(-11)$

The absolute value of –5 is 5.
The absolute value of –11 is 11.
Add the absolute values.

  $5 + 11 = 16$

Write a negative sign in front of the sum.

  $-5+(-11)=-16$

b. $-8+(-32)$

Sum of the absolute values, with a negative sign written in front.

  $-8+(-32)=-40$

c. $-21+(-62)$

Sum of the absolute values, with a negative sign written in front.

  $-21+(-62)=-83$

d. $-\dfrac{3}{5}+\left(-\dfrac{2}{3}\right)$

The absolute value of $-\dfrac{3}{5}$ is $\dfrac{3}{5}$, and the absolute

value of $-\dfrac{2}{3}$ is $\dfrac{2}{3}$. Add the absolute values.

Check that the sum is in lowest terms.

  $$\frac{3}{5}+\frac{2}{3}=\frac{9}{15}+\frac{10}{15}=\frac{19}{15}$$

Write a negative sign in front of the sum.

  $$-\frac{3}{5}+\left(-\frac{2}{3}\right)=-\frac{19}{15}$$

Now Try:

2. Add (without using number lines).

a. $-7+(-12)$

_____

b. $-3+(-39)$

_____

c. $-49+(-59)$

_____

d. $-\dfrac{5}{6}+\left(-\dfrac{2}{3}\right)$

_____

3. Add each sum.

a. $7+(-2)$

First find this sum with a number line.

$$\frac{-2}{7}$$

Because the second arrow ends at 5,

$7+(-2)=5$

Now find the sum using the rule. First, find the absolute value of each number.

$|7|=7 \qquad |-2|=2$

Subtract the lesser absolute value from the greater absolute value.

$7-2=5$

The positive number 7 has the greater absolute value, so the answer is positive.

$7+(-2)=5$

b. $5+(-13)$

First, find the absolute values.

$|5|=5 \qquad |-13|=13$

Subtract the lesser absolute value from the greater absolute value.

$13-5=8$

The negative number –13 has the greater absolute value, so the answer is negative.

$5+(-13)=-8$

Write a negative sign in front of the answer because –13 has the greater absolute value.

c. $-24+19$

$-24+19=-5$

Write a negative sign in front of the answer because –24 has the greater absolute value.

d. $14+(-8)$

$14+(-8)=6$

Positive answer because the positive number 14 has the greater absolute value.

3. Add each sum.

a. $10+(-7)$

_____

b. $5+(-17)$

_____

c. $-29+14$

_____

d. $23+(-9)$

_____

e. $-\dfrac{1}{3}+\dfrac{1}{4}$

The absolute value of $-\dfrac{1}{3}$ is $\dfrac{1}{3}$, and the absolute

value of $\dfrac{1}{4}$ is $\dfrac{1}{4}$. Subtract the lesser absolute

value from the greater absolute value.

$$\dfrac{1}{3}-\dfrac{1}{4}=\dfrac{4}{12}-\dfrac{3}{12}=\dfrac{1}{12}$$

Because the negative number $-\dfrac{1}{3}$ has the greater

absolute value, the answer is negative.

$$-\dfrac{1}{3}+\dfrac{1}{4}=-\dfrac{1}{12}$$

e. $-\dfrac{1}{10}+\dfrac{2}{15}$

_____

Review these examples for Objective 3:

4. Give the additive inverse of each number. Then find the sum of the number and its inverse.

a. 5

Additive inverse: –5
Sum of Number and Inverse: $5+(-5)=0$

b. –9

Additive inverse: $-(-9)$ or 9
Sum of Number and Inverse: $9+(-9)=0$

c. 11

Additive inverse: –11
Sum of Number and Inverse: $11+(-11)=0$

d. –2

Additive inverse: $-(-2)$ or 2
Sum of Number and Inverse: $-2+2=0$

e. $\dfrac{5}{7}$

Additive inverse: $-\dfrac{5}{7}$

Sum of Number and Inverse: $\dfrac{5}{7}+\left(-\dfrac{5}{7}\right)=0$

f. 0

Additive inverse: 0
Sum of Number and Inverse: $0+0=0$

Now Try:

4. Give the additive inverse of each number. Then find the sum of the number and its inverse.

a. 8

_____

b. –25

_____

c. 6.9

_____

d. –300

_____

e. $-\dfrac{8}{5}$

_____

f. 0

_____

Review these examples for Objective 4:

5. Subtract.

a. $9 - 12$

The first number, 9, stays the same. Change the subtraction sign to addition. Change the sign of the second number to its opposite.

$$9 - \quad 12$$

$$9 + (-12)$$

Now add.

$$9 + (-12) = -3$$

So $9 - \quad 12 = -3$ also.

b. $-7 - 14$

Subtraction is changed to addition. Positive 14 is changed to its opposite, (-14).

$$-7 - \quad 14$$

$$-7 + (-14) = -21$$

c. $-6 - (-8)$

Subtraction is changed to addition. Negative 8 is changed to its opposite, $(+8)$.

$$-6 - (-8)$$

$$-6 + (+8) = 2$$

d. $12 - (-5)$

Subtraction is changed to addition. Negative 5 is changed to its opposite, $(+5)$.

$$12 - (-5)$$

$$12 + (+5) = 17$$

e. $-35 - (-31)$

Subtraction is changed to addition. Negative 31 is changed to its opposite, $(+31)$.

$$-35 - (-31)$$

$$-35 + (+31) = -4$$

Now Try:

5. Subtract.

a. $12 - 17$

_____

b. $-8 - 24$

_____

c. $-9 - (-14)$

_____

d. $13 - (-8)$

_____

e. $-41 - (-39)$

_____

6. Subtract.

a. $3.5-(-9.4)$

Subtraction is changed to addition. Negative 9.4 is changed to its opposite, (+9.4).

$3.5+(+9.4)=12.9$

Once the subtraction is changed to addition, notice that both numbers are positive.

$$\begin{array}{r} 3.5 \\ +\ 9.4 \\ \hline 12.9 \end{array}$$

b. $-0.4-13.2$

Subtraction is changed to addition. Positive 13.2 is changed to its opposite, (–13.2).

$-0.4+(-13.2)=-13.6$

To add two negative numbers, first add the absolute values.

$|-0.4|$ is 0.4 $|-13.2|$ is 13.2

$$\begin{array}{r} 0.4 \\ +\ 13.2 \\ \hline 13.6 \end{array}$$

Then write a negative sign in front of 13.6

$-0.4+(-13.2)=-13.6$

c. $\dfrac{2}{3}-\dfrac{5}{6}$

Subtraction is changed to addition. Positive $\dfrac{5}{6}$ is changed to its opposite, $\left(-\dfrac{5}{6}\right)$.

$$\frac{2}{3}+\left(-\frac{5}{6}\right)=\frac{4}{6}+\left(-\frac{5}{6}\right)=-\frac{1}{6}$$

Once the subtraction is changed to addition, notice that the numbers have different signs. So we subtract the lesser absolute value from the greater absolute value.

$\left|\dfrac{4}{6}\right|$ is $\dfrac{4}{6}$ and $\left|-\dfrac{5}{6}\right|$ is $\dfrac{5}{6}$ Then $\dfrac{5}{6}-\dfrac{4}{6}=\dfrac{1}{6}$

The negative number $-\dfrac{5}{6}$ has the greater absolute value, so the answer is negative.

6. Subtract.

a. $13.6-(-11.5)$

_____

b. $-5.7-12.8$

_____

c. $\dfrac{1}{5}-\dfrac{3}{4}$

_____

d. $-\dfrac{3}{4}-\left(-\dfrac{3}{10}\right)$

Subtraction is changed to addition. Negative $\dfrac{3}{10}$

is changed to its opposite, $\left(-\dfrac{3}{10}\right)$.

$$-\dfrac{3}{4}-\left(-\dfrac{3}{10}\right)=-\dfrac{15}{20}+\left(+\dfrac{6}{20}\right)=-\dfrac{9}{20}$$

Once the subtraction is changed to addition, notice that the numbers have different signs. So we subtract the lesser absolute value from the greater absolute value.

$\left|-\dfrac{15}{20}\right|$ is $\dfrac{15}{20}$ and $\left|+\dfrac{6}{20}\right|$ is $\dfrac{6}{20}$ Then $\dfrac{15}{20}-\dfrac{6}{20}$ is $\dfrac{9}{20}$

The negative number $-\dfrac{15}{20}$ has the greater

absolute value so the answer is negative.

d. $-\dfrac{7}{8}-\left(-\dfrac{5}{6}\right)$

_____

Review these examples for Objective 5:

7. According to the last step in the order of operations, perform addition and subtraction from left to right.

a. $-8+(-12)-7$

Change subtraction to addition; change positive 7 to its opposite, (–7).

$$-8+(-12)-7$$
$$-20\quad -7$$
$$-20+(-7)$$
$$-27$$

b. $8-(-6)+(-7)$

Change subtraction to addition; change –6 to its opposite, (+6).

$$8-(-6)+(-7)$$
$$8+(+6)+(-7)$$
$$14\quad +(-7)$$
$$7$$

Now Try:

7. According to the last step in the order of operations, perform addition and subtraction from left to right.

a. $-5+(-17)-3$

_____

b. $9-(-8)+(-6)$

_____

c. $15+(-4)-(-7)+14$

Change subtraction to addition; change -7 to its opposite, $(+7)$.

$$15+(-4)-(-7)+14$$

$$11 \quad\quad -(-7)+14$$

$$11 \quad\quad +(+7)+14$$

$$18 \quad\quad\quad +14$$

$$32$$

d. -7.2
   -15.6
   9.3
   -6.5
   $\underline{4.9}$

Start at the top.
   -7.2
   $-15.6 \quad\quad -22.8$
   $9.3 \quad\quad\quad 9.3 \quad\quad -13.5$
   $-6.5 \quad\quad -6.5 \quad\quad -6.5 \quad\quad -20.0$
   $\underline{4.9} \quad\quad \underline{4.9} \quad\quad \underline{4.9} \quad\quad \underline{4.9}$
   $\quad\quad\quad\quad\quad\quad\quad\quad\quad\quad\quad\quad -15.1$

c. $12+(-7)-(-3)+9$

_____

d. -13.5
   -7.3
   9.6
   -3.8
   $\underline{5.7}$

_____

Objective 1 Add signed numbers by using a number line.

For extra help, see Example 1 on pages 632–633 of your text and Section Lecture video for Section 9.2.

Add by using a number line.

1. $8 + 7$

1. _____

2. $7 + (-12)$

2. _____

3. $-2 + (-6)$

3. _____

Objective 2   Add signed numbers without using a number line.

For extra help, see Examples 2–3 on pages 633–635 of your text and Section Lecture video for Section 9.2 and Exercise Solutions Clip 11 and 15.

Add.

4. $-10+17$ 4. _____

5. $-3.2+(-4.7)$ 5. _____

6. $-\dfrac{1}{2}+\dfrac{5}{4}$ 6. _____

Objective 3   Find the additive inverse of a number.

For extra help, see Examples 3–4 on page 636 of your text and Section Lecture video for Section 9.2.

Give the additive inverse of each number.

7. -15 7. _____

8. 0 8. _____

9. 281 9. _____

Objective 4   Subtract signed numbers.

For extra help, see Examples 5–6 on pages 637–638 of your text and Section Lecture video for Section 9.2 and Exercise Solutions Clip 51.

Subtract.

10. $-1-(-5)$ 10. _____

11. $-\dfrac{1}{2}-\left(-\dfrac{3}{8}\right)$ 11. _____

12. $4.9-(-8.3)$ 12. _____

Objective 5   Add or subtract a series of signed numbers.

For extra help, see Example 7 on page 639 of your text and Section Lecture video for Section 9.2.

Follow the order of operations to work each problem.

13. $-6-(-1)+(-9)$ 13. _____

14. $-4.8-(-3.6)+6.4$ 14. _____

15. $6.8+(-5.9)-(-8.6)$ 15. _____

Chapter 9 BASIC ALGEBRA

9.3   Multiplying and Dividing Signed Numbers

Learning Objectives
1   Multiply or divide two numbers with opposite signs.
2   Multiply or divide two numbers with the same sign.

Key Terms

Use the vocabulary terms listed below to complete each statement in exercises 1–3.

factors   **product**   **quotient**

1.   Numbers that are being multiplied are called _____.

2.   The answer to a division problem is called the _____.

3.   The answer to a multiplication problem is called the _____.

Guided Examples

Review these examples for Objective 1:

1.   Multiply.

   a. $-7 \cdot 5$

   Factors have different signs, so the product is negative.
   $$-7 \cdot 5 = -35$$

   b. $8(-7)$

   Factors have different signs, so the product is negative.
   $$8(-7) = -56$$

   c. $(-9)(12)$

   $$(-9)(12) = -108$$

   d. $14(-7)$

   $$14(-7) = -98$$

Now Try:

1.   Multiply.

   a. $-14 \cdot 5$

   _____

   b. $7(-9)$

   _____

   c. $(-6)(12)$

   _____

   d. $22(-8)$

   _____

Name: Date:

Instructor: Section:

Review these examples for Objective 2: | **Now Try:**
2. Multiply. | 2.   Multiply.

    a. $(-8)(-3)$ | **a.** $(-4)(-13)$

The factors have the same sign (both are negative). The product is positive.

$$(-8)(-3) = 24$$

    b. $-9(-5)$ | **b.** $-15(-6)$

$$-9(-5) = 45$$

    c. $(-10)(-8)$ | **c.** $(-8)(-9)$

$$(-10)(-8) = 80$$

    d. $(-20)(-6)$ | **d.** $(-15)(-1)$

$$(-20)(-6) = 120$$

    e. $8(7)$ | **e.** $12(7)$

$$8(7) = 56$$

3. Divide. | 3.   Divide.

    a. $\dfrac{-18}{6}$ | **a.** $\dfrac{-24}{8}$

The numbers have different signs, so the quotient is negative.

$$\frac{-18}{6} = -3$$

    b. $\dfrac{-9}{-3}$ | **b.** $\dfrac{-16}{-4}$

The numbers have the same sign (both negative), so the quotient is positive.

$$\frac{-9}{-3} = 3$$

    c. $\dfrac{-80}{-20}$ | **c.** $\dfrac{-125}{-5}$

$$\frac{-80}{-20} = 4$$

d. $\dfrac{-5}{0}$

Division by 0 cannot be done.

$\dfrac{-5}{0}$ is undefined.

e. $\dfrac{0}{-7}$

0 can be divided by a nonzero number.

$\dfrac{0}{-7}=0$

f. $\dfrac{99}{-11}$

$\dfrac{99}{-11}=-9$

g. $\dfrac{-\frac{3}{4}}{-\frac{5}{8}}$

Use the reciprocal of $-\dfrac{5}{8}$, which is $-\dfrac{8}{5}$. Then divide out the common factor and multiply.

$\dfrac{-\frac{3}{4}}{-\frac{5}{8}}=\left(-\dfrac{3}{4}\right)\left(-\dfrac{8}{5}\right)$

$=\left(-\dfrac{3}{\cancel{4}_1}\right)\left(-\dfrac{\cancel{8}^2}{5}\right)$

$=\dfrac{6}{5}$

d. $\dfrac{-10}{0}$

_____

e. $\dfrac{0}{-13}$

_____

f. $\dfrac{120}{-40}$

_____

g. $\dfrac{-\frac{4}{5}}{-\frac{8}{15}}$

_____

Objective 1   Multiply or divide two numbers with opposite signs.

For extra help, see Example 1 on page 644 of your text and Section Lecture video for Section 9.3 and Exercise Solutions Clip 5, 27, 31, 55, and 81.

Multiply or divide as indicated.

1. $(-13)(3)$ **1.** _____

2. $-96 \div 12$ 2. _____

3. $5 \div \left(-\dfrac{5}{11}\right)$ 3. _____

Objective 2   Multiply or divide two numbers with the same sign.

For extra help, see Examples 2–3 on page 645 of your text and Section Lecture video for Section 7.1 and Exercise Solutions Clip 13 and 69.

Multiply or divide as indicated.

4. $-\dfrac{7}{10} \cdot \left(-\dfrac{5}{4}\right)$ 4. _____

5. $\dfrac{22.75}{5}$ 5. _____

6. $\dfrac{-\frac{9}{20}}{-\frac{3}{4}}$ 6. _____

Chapter 9 BASIC ALGEBRA

9.4   Order of Operations

Learning Objectives
1   Use the order of operations.
2   Use the order of operations with exponents.
3   Use the order of operations with fraction bars.

Key Terms

Use the vocabulary terms listed below to complete each statement in exercises 1–3.

exponent   base   order of operations

1.   For problems or expressions with more than one operation, the
_____ tells what to do first, second, and so on, to
obtain the correct answer.

2.   In the expression 3^5, the _____ is 3.

3.   In the expression 3^5, the _____ is 5.

Guided Examples

Review these examples for Objective 1:

1.   Use the order of operations to simplify this expression.
$$2 - 12 \div 3 + 4$$

Check for parentheses: none.
Check for exponents and square roots: none.

$2 - 12 \div 3 + 4$ Divide from left to right.

$2 - \quad 4 \ + 4$ Change 4 to its opposite (-4).

$2 + (-4) \ + 4$ Add $2 + (-4)$.

$-2 \qquad + 4$ Add $-2 + 4$.

2

Now Try:

1.   Use the order of operations to
simplify this expression.
$$5 - 21 \div 3 + 6$$

_____

Name: Date:
Instructor: Section:

2. Use the order of operations to simplify each expression.

    a. $-6(8-3)-4$

Work inside parentheses first.
$-6(8-3)-4$

$-6(5)\quad -4$ Check for exponents and
                   square roots: none

$-6(5)\quad -4$ Multiply from left to right.

$-30\quad\quad -4$ Change subtraction to addition.

$-30\;+(-4)$ Add.

    -34

    b. $8+3(7-11)(18\div 6)$

Work inside first set of parentheses.
$8+3(7-11)(18\div 6)$ Change $7-11$ to
                    $7+(-11)$ to get -4.

$8+\;\;3(-4)(18\div 6)$ Divide $18\div 6 = 3$.

$8+\;\;3(-4)\;\;(3)$ Multiply from left to right.

$8+\;\;(-12)\;\;(3)$ Multiply $(-12)(3)=-36$.

$8+\;\;\;\;(-36)$ Add.

    -28

2. Use the order of operations to simplify each expression.

    a. $-7(9-5)-8$

    _____

    b. $5+6(7-9)(24\div 8)$

    _____

Review these examples for Objective 2:

3. Simplify.

    a. $6^2-(-5)^2$

There are parentheses around (–5), but no work can be done inside these parentheses. Apply the exponents, then subtract.

$6^2-(-5)^2$

$36-\;25$

   11

Now Try:

3. Simplify.

    a. $7^2-(-6)^2$

    _____

b. $(-8)^2 - (7-9)^2(-5)$

Work inside the parentheses first.

$(-8)^2 - (7-9)^2(-5)$

$(-8)^2 - (-2)^2(-5)$ Apply the exponents.

$64 \quad - \quad 4(-5)$ Multiply.

$64 \quad -(-20)$ Change subtraction to addition.

$64 + (+20)$ Add.

$\quad 84$

c. $\left(\dfrac{3}{5} - \dfrac{1}{10}\right)^2 \div \left(-\dfrac{5}{8}\right)$

Work inside parentheses first.

$\dfrac{3}{5} - \dfrac{1}{10} = \dfrac{6}{10} - \dfrac{1}{10} = \dfrac{5}{10} = \dfrac{1}{2}$

$\left(\dfrac{3}{5} - \dfrac{1}{10}\right)^2 \div \left(-\dfrac{5}{8}\right)$

$\left(\dfrac{1}{2}\right)^2 \div \left(-\dfrac{5}{8}\right)$ Apply the exponent.

$\dfrac{1}{4} \quad \div \left(-\dfrac{5}{8}\right)$ Divide by using the reciprocal of the divisor.

$\dfrac{1}{4} \quad \cdot \left(-\dfrac{8}{5}\right)$ Divide out common factors, then multiply: $\dfrac{1}{\overset{}{\underset{1}{\cancel{4}}}} \cdot -\dfrac{\overset{2}{\cancel{8}}}{5} = -\dfrac{2}{5}.$

$-\dfrac{2}{5}$

b. $(-9)^2 - (-4)^2(-3)$

_____

c. $\left(\dfrac{3}{4} - \dfrac{5}{8}\right)^2 \div \left(-\dfrac{5}{16}\right)$

_____

Review this example for Objective 3:
4. Simplify.

$$\dfrac{-7+6(5-8)}{9-6^2 \div 9}$$

First do work in the numerator.

$-7+6(5-8)$ Work inside parentheses.

$-7+6(-3)$ Multiply.

$-7+(-18)$ Add.

$-25 \leftarrow$ Numerator.

Now do work in the denominator.

$9-6^2 \div 9$

Now Try:
4. Simplify.

$$\dfrac{-8+4(6-9)}{9-8^2 \div 16}$$

_____

$9 - 6^2 \div 9$ Apply the exponent.

$9 - 36 \div 9$ Divide.

$9 - \quad 4$ Subtract.

$\quad 5 \leftarrow$ Denominator.

The last step is the division.

$$\text{Numerator} \rightarrow \frac{-25}{5} = -5$$
$$\text{Denominator} \rightarrow$$

Objective 1   Use the order of operations.

For extra help, see Examples 1–2 on pages 650–651 of your text and Section Lecture video for Section 9.4 and Exercise Solutions Clip 21 and 33.

Simplify.

1. $4 + (-3) + 2 \cdot (-5)$ 1. _____

2. $8 \div (-4) + 4 \cdot (-3) + 2$ 2. _____

3. $-3 \cdot (8 - 16) \div (-8)$ 3. _____

Objective 2   Use the order of operations with exponents.

For extra help, see Example 3 on page 652 of your text and Section Lecture video for Section 9.4 and Exercise Solutions Clip 9, 11, 17, and 27.

Simplify.

4. $3 \cdot 5^2 - 3 \cdot 7 - 9$ 4. _____

5. $-\left(\frac{3}{7} + \frac{2}{7}\right) + \left(-\frac{1}{3}\right)^2$ 5. _____

6. $-\left(-\frac{1}{6} + \frac{5}{6}\right) \div \left(-\frac{1}{3}\right)^2$ 6. _____

Name: Date:

Instructor: Section:

Objective 3   Use the order of operations with fraction bars.

For extra help, see Example 4 on page 653 of your text and Section Lecture video for Section 9.4 and Exercise Solutions Clip 41.

Simplify.

7. $\dfrac{4^3 - 3^3}{-5(-4+2)}$ 7. _____

8. $\dfrac{5(-8+3)}{13(-2)+(-6-1)(-4+1)}$ 8. _____

9. $\dfrac{-8-(-5-7)}{5-(-3)^2}$ 9. _____

Chapter 9 BASIC ALGEBRA

9.5   Evaluating Expressions and Formulas

Learning Objectives
1   Define variable and expression.
2   Evaluate an expression when values of the variables are given.

Key Terms

Use the vocabulary terms listed below to complete each statement in exercises 1–2.

    variable   **expression**

1.   A combination of operations on variables and numbers is called an

_____ .

2.   A letter that represents a number is called a _____ .

Guided Examples

Review these examples for Objective 1:

1.   Evaluate $4x - 5y$, if x is 3 and y is 6.

Replace x with 3. Replace y with 6. Then use the order of operations.

$$4\,x - 5\,y$$

$$4(3) - 5(6) \quad \text{Multiply.}$$

$$12 - 30 \quad \text{Subtract.}$$

$$-18$$

2.   Evaluate $\dfrac{8k + 3r}{4s}$, if k is –3, r is 6 and s is –2.

Replace k with –3, r with 6, and s with –2.

$$\frac{8k + 3r}{4s} = \frac{8(-3) + 3(6)}{4(-2)} \quad \text{Do multiplication first.}$$

$$= \frac{-24 + 18}{-8} \quad \text{Add in the numerator.}$$

$$= \frac{-6}{-8}$$

$$= \frac{3}{4}$$

Now Try:

1.   Evaluate $8x - 7y$, if x is 4 and y is 5.

_____

2.   Evaluate $\dfrac{7k + 4r}{8s}$, if k is –4, r is 5 and s is –1.

_____

3. Evaluate $-c - 4b$, when c is -5 and b is -2.

Replace c with -5. Replace b with -2.

$-c \quad -4 \quad b$

$-(-5) - 4(-2)$ Multiply $4 \cdot (-2)$.

$+5 - (-8)$ Change subtraction to addition; change (-8) to its opposite, $(+8)$.

$5 + (+8)$

13

3. Evaluate $-x - 7y$, when x is -8 and y is -1.

_____

4. The area of a triangle is $A = \frac{1}{2}bh$. In this formula, b is the length of the base of the triangle and h is the height of the triangle. What is the area if b is 8 m and h is 9 m?

Replace b with 8 m and h with 9 m.

$A = \frac{1}{2} \quad b \quad h$

$A = \frac{1}{2}(8 \text{ m})(9 \text{ m})$

$A = \frac{1}{\cancel{2}}(\overset{4}{\cancel{8}} \text{ m})(9 \text{ m})$ Divide out any common factors.

$A = 36 \text{ m}^2$

The area of the triangle is 36 m^2.

4. The area of a triangle is $A = \frac{1}{2}bh$. In this formula, b is the length of the base of the triangle and h is the height of the triangle. What is the area if b is 7 ft and h is 18 ft?

_____

Objective 1   Define variable and expression.

For extra help, see page 660 of your text and Section Lecture video for Section 9.5.

*Write if each of the following is a **variable** or an **expression**.*

1. $3p$

2. r

3. $7x + 4y$

1. _____

2. _____

3. _____

Objective 2   Evaluate an expression when values of the variables are given.

For extra help, see Examples 1–4 on pages 660–661 of your text and Section Lecture video for Section 9.5 and Exercise Solutions Clip 13, 15, 17, 23, and 37.

Find the value of the expression $3r - 2s$ *for the following values of r and s.*

4. $r = 0,\ s = -12$ 4. _____

Use the given values of the variables to find the value of the expression.

5. $-4k + 3m;\ k = 5,\ m = -\dfrac{1}{3}$ 5. _____

Using the given values, evaluate the formula. Round to the nearest hundredth, if necessary.

6. $s = 2\pi r^2 + 2\pi rh;\ \pi = 3.14,\ r = 3,\ h = 5$ 6. _____

Chapter 9 BASIC ALGEBRA

9.6   Solving Equations

Learning Objectives
1   Determine whether a number is a solution of an equation.
2   Solve equations using the addition property of equality.
3   Solve equations using the multiplication property of equality.

Key Terms

Use the vocabulary terms listed below to complete each statement in exercises 1–4.

equation   **solution**   **addition property of equality**

multiplication property of equality

1.   An _____ is a statement that says two expressions are equal.

2.   The _____ states that both sides of an equation can be multiplied or divided by the same number, except division by 0 cannot be done.

3.   The _____ of an equation is a number that can replace the variable so that the equation is true.

4.   The _____ states that the same number can be added to or subtracted on both sides of an equation.

Guided Examples

Review these examples for Objective 1:

1.   Is 8 a solution of either one of these equations?

    a. $17 = x + 9$

Replace x with 8.
$$17 = x + 9$$
$$17 = 8 + 9$$
$$17 = \ 17 \quad \text{True}$$
Because the statement is true, 8 is a solution of the equation $17 = x + 9$.

    b. $3y + 4 = 35$

Replace y with 8.
$$3y + 4 = 35$$
$$3(8) + 4 = 35$$
$$24 + 4 = 35$$
$$28 \neq 35 \quad \text{False}$$
The false statement shows that 8 is not a solution of $3y + 4 = 35$.

Now Try:

1.   Is 6 a solution of either one of these equations?

    a. $13 = p + 7$

        _____

    b. $35 = 5r$

        _____

Name: Date:
Instructor: Section:

Review these examples for Objective 2:

2. Solve each equation.

    a. $k-6=10$

    To get k by itself on the left side, add 6 to the left side because $k-6+6$ gives $k+0$. You must then add 6 to the right side also.
    $$k-6=10$$
    $$k-6+6=10+6$$
    $$k+0 \;\; =16$$
    $$k \quad\;\; =16$$
    The solution is 16. To check the solution, replace k with 16 in the original equation.
    $$k-6=10 \leftarrow \text{Original equation}$$
    $$16-6=10 \quad \text{Replace } k \text{ with 16.}$$
    $$10=10 \quad \text{True}$$
    The result is true, so 16 is the solution.

    b. $4=z+9$

    To get z by itself on the right side, add -9 to both sides.
    $$4=z+9$$
    $$4+(-9)=z+9+(-9)$$
    $$-5=z+0$$
    $$-5=z$$
    Check the solution by replacing z with -5 in the original equation.
    $$4=z+9$$
    $$4=z+9 \leftarrow \text{Original equation}$$
    $$4=-5+9 \quad \text{Replace } z \text{ with } -5.$$
    $$4=4 \qquad\qquad \text{True}$$
    The result is true, so -5 is the solution.

Review these examples for Objective 3:

3. Solve each equation.

    a. $8p=72$

    You want to get the variable, p, by itself on the left side. The expression $8p$ means $8 \cdot p$. To undo the multiplication and get p by itself, divide both sides by 8.

Now Try:

2. Solve each equation.

    a. $n-6=10$

    _____

    b. $8=x+12$

    _____

Now Try:

3. Solve each equation.

    a. $9p=45$

    _____

$$8p = 72$$

$$\frac{\overset{1}{\cancel{8}} \cdot p}{\underset{1}{\cancel{8}}} = \frac{72}{8} \quad \text{Divide both sides by 8.}$$

$$p = 9 \quad \text{The solution is 9.}$$

Check

$$8p = 72 \leftarrow \text{Original equation}$$

$$8(9) = 72 \quad \text{Replace } p \text{ with 9.}$$

$$72 = 72 \quad \text{True}$$

The result is true, so 9 is the solution.

b. $-6r = 30$

Divide both sides by –6 to get r by itself on the left side.

$$\frac{\overset{1}{\cancel{-6}} \cdot r}{\underset{1}{\cancel{-6}}} = \frac{30}{-6} \quad \text{Divide both sides by } -6.$$

$$r = -5 \quad \text{The solution is } -5.$$

Check

$$-6r = 30 \leftarrow \text{Original equation}$$

$$-6(-5) = 30 \quad \text{Replace } r \text{ with } -5.$$

$$30 = 30 \quad \text{True}$$

The result is true, so –5 is the solution (not 30).

c. $-72 = -12m$

Divide both sides by –12 to get m by itself on the right side.

$$\frac{-72}{-12} = \frac{\overset{1}{\cancel{-12}} \cdot m}{\underset{1}{\cancel{-12}}} \quad \text{Divide both sides by } -12.$$

$$6 = m \quad \text{The solution is 6.}$$

Check

$$-72 = -12m \leftarrow \text{Original equation}$$

$$-72 = -12(6) \quad \text{Replace } m \text{ with 6.}$$

$$-72 = -72 \quad \text{True}$$

The result is true, so 6 is the solution (not –72).

b. $-7r = 28$

_____

c. $-64 = -16m$

_____

4. Solve each equation.

a. $\dfrac{x}{3} = 8$

Rewrite $\dfrac{x}{3}$ as $\dfrac{1}{3}x$, because dividing x by 3

is the same as multiplying x by $\dfrac{1}{3}$. Then, to

get x by itself, multiply both sides by the

reciprocal of $\dfrac{1}{3}$, which is $\dfrac{3}{1}$.

$$\dfrac{1}{3}x = 8$$

$$\dfrac{\cancel{3}}{1} \cdot \dfrac{1}{\cancel{3}}x = \dfrac{3}{1} \cdot 8$$

$$1x = 24$$

$$x = 24 \quad \text{The solution is 24.}$$

Check

$$\dfrac{x}{3} = 8 \leftarrow \text{Original equation}$$

$$\dfrac{24}{3} = 8 \quad \text{Replace } x \text{ with 24.}$$

$$8 = 8 \quad \text{True}$$

24 is the correct solution (not 8).

b. $-\dfrac{3}{4}r = 15$

Multiply both sides by the reciprocal of

$-\dfrac{3}{4}$, which is $-\dfrac{4}{3}$.

$$-\dfrac{3}{4}r = 15$$

$$-\dfrac{\cancel{4}}{\cancel{3}} \cdot \left(-\dfrac{\cancel{3}}{\cancel{4}}r\right) = -\dfrac{4}{\cancel{3}} \cdot \dfrac{\cancel{15}}{1}$$

$$r = -20 \quad \text{The solution is } -20.$$

Check by replacing r with -20 in the original

equation. Write -20 as $\dfrac{-20}{1}$.

4. Solve each equation.

a. $\dfrac{x}{5} = 7$

_____

b. $-\dfrac{5}{6}r = 25$

_____

$$-\frac{3}{4}r = 15 \leftarrow \text{Original equation}$$

$$-\frac{3}{\cancel{4}_1} \cdot \frac{\cancel{-20}^{-5}}{1} = 15 \quad \text{Replace } r \text{ with } -20.$$

$$15 = 15 \quad \text{True}$$

-20 is the correct solution (not 15).

Objective 1   Determine whether a number is a solution of an equation.

For extra help, see Example 1 on page 665 of your text and Section Lecture video for Section 9.6 and Exercise Solutions Clip 5.

Decide whether the given number is a solution of the equation.

1. $b - 5 = 18;\ 13$ 1. _____

2. $5 + 8m = 3;\ -1$ 2. _____

3. $-5y + 1 = 6;\ -1$ 3. _____

Objective 2   Solve equations using the addition property of equality.

For extra help, see Example 2 on pages 666–667 of your text and Section Lecture video for Section 9.6 and Exercise Solutions Clip 23 and 33.

Solve each equation using the addition property. Check each solution.

4. $y + 11 = 16$ 4. _____

5. $-12 = -10 + a$ 5. _____

6. $x - 1.24 = 4.37$ 6. _____

Objective 3 Solve equations using the multiplication property of equality.

For extra help, see Examples 3–4 on pages 667–669 of your text and Section Lecture video for Section 9.6 and Exercise Solutions Clip 43, 49, and 63.

Solve each equation using the multiplication property. Check each solution.

7. $1.32 = -1.2m$

7. _____

8. $-12 = \dfrac{r}{3}$

8. _____

9. $-\dfrac{1}{4}x = 8$

9. _____

Chapter 9 BASIC ALGEBRA

9.7   Solving Equations with Several Steps

Learning Objectives
1   Solve equations with several steps.
2   Use the distributive property.
3   Combine like terms.
4   Solve more difficult equations.

Key Terms

Use the vocabulary terms listed below to complete each statement in exercises 1–2.

> **distributive property**   **like terms**

1.   Terms that have exactly the same variables and exponents are called

   _____.

2.   The _____ states that $a(b + c) = ab + bc$.

Guided Examples

Review this example for Objective 1:

1.   Solve $6m + 2 = 32$.

   Step 1 Subtract 2 from both sides so that $6m$ will be by itself on the left side.
   $$6m + 2 - 2 = 32 - 2$$
   $$6m = 30$$

   Step 2 Divide both sides by 6.
   $$\frac{\overset{1}{\cancel{6}} \cdot m}{\underset{1}{\cancel{6}}} = \frac{30}{6}$$

   $$m = 5 \quad \text{The solution is 5.}$$

   Step 3 Check the solution.
   $$6m + 2 = 32 \leftarrow \text{Original equation}$$
   $$6(5) + 2 = 32 \quad \text{Replace } m \text{ with 5.}$$
   $$30 + 2 = 32$$
   $$32 = 32 \quad \text{True}$$

   5 is the correct solution (not 32).

Now Try:

1.   Solve $-7 = 3p + 5$.

   _____

Review these examples for Objective 2:

2. Simplify each expression by using the distributive property.

   a. $8(3+5)$

   $8(3+5)=8\cdot3+8\cdot5=24+40=64$
   The 8 outside the parentheses is distributed over the 3 and 5 inside the parentheses. That means that every number inside the parentheses is multiplied by 8.

   b. $-4(k+7)$

   $-4(k+7)=-4\cdot k+(-4)\cdot7=-4k+(-28)=-4k-28$

   c. $5(y-9)$

   $5(y-9)=5\cdot y-5\cdot9=5y-45$

   d. $-6(x-2)$

   $-6(x-2)=-6\cdot x-6(-2)=-6x-(-12)=-6x+12$

Now Try:

2. Simplify each expression by using the distributive property.

   a. $9(6+7)$

   _____

   b. $-2(k+5)$

   _____

   c. $7(y-8)$

   _____

   d. $-8(x-4)$

   _____

Review these examples for Objective 3:

3. Use the distributive property to combine like terms.

   a. $6k+13k$

   $6k+13k=(6+13)k=19k$

   b. $11m-17m+3m$

   $11m-17m+3m=(11-17+3)m=-3m$

   c. $-7x+x$

   $-7x+x=-7x+1x=(-7+1)x=-6x$

Now Try:

3. Use the distributive property to combine like terms.

   a. $9k+14k$

   _____

   b. $23m-29m+4m$

   _____

   c. $-12x+x$

   _____

Review these examples for Objective 4:

4. Solve each equation. Check each solution.

   a. $5r+4r=45$

   You can combine $5r$ and $4r$ because they are like terms. $5r+4r$ is $9r$, so the equation becomes

Now Try:

4. Solve each equation. Check each solution.

   a. $5y-2=3y+8$

   _____

$$9r = 45$$

$$\frac{\cancel{9} \cdot r}{\cancel{9}} = \frac{45}{9} \quad \text{Divide both sides by 9.}$$

$$r = 5$$

Check

$$5r + 4r = 45 \leftarrow \text{Original equation}$$

$$5(5) + 4(5) = 45 \qquad \text{Replace } r \text{ with 5.}$$

$$25 + 20 = 45$$

$$45 = 45 \qquad \text{True}$$

5 is the correct solution (not 45).

b. $3k + 6 = 9k - 24$

One way to get the variable term by itself on one side is to subtract $9k$ from both sides.

$$3k + 6 - 9k = 9k - 24 - 9k$$

$$3k - 9k + 6 = 9k - 9k - 24$$

$$-6k + 6 = -24$$

Next, subtract 6 from both sides.

$$-6k + 6 - 6 = -24 - 6$$

$$-6k = -30$$

Finally, divide both sides by –6.

$$\frac{\cancel{-6} \cdot k}{\cancel{-6}} = \frac{-30}{-6}$$

$$k = 5$$

Check

$$3k + 6 = 9k - 24 \leftarrow \text{Original equation}$$

$$3(5) + 6 = 9(5) - 24 \quad \text{Replace } k \text{ with 5.}$$

$$15 + 6 = 45 - 24$$

$$21 = 21 \qquad \text{True}$$

5 is the correct solution (not 21).

5. Solve $-20 = 4(y - 5)$.

Step 1 Use the distributive property on the right side of the equation.

$4(y - 5)$ becomes $4 \cdot y - 4 \cdot 5$ or $4y - 20$.

Now the equation looks like this.

$$-20 = 4y - 20$$

Step 2 Combine like terms. Check the left side

b. $4p - 8 = 7p - 20$

_____

5. Solve $7(m + 3) = 28$.

_____

of the equation. There are no like terms. Check the right side. No like terms there either, so go on to Step 3.

Step 3 Add 20 to both sides to get the variable term by itself on the right side.

$$-20 + 20 = 4y - 20 + 20$$

$$0 = 4y$$

Step 4 Divide both sides by 4.

$$\frac{0}{4} = \frac{\overset{1}{\cancel{4}} \cdot y}{\underset{1}{\cancel{4}}}$$

$$0 = y$$

Step 5 Check. Go back to the original equation.

$$-20 = 4(y - 5) \leftarrow \text{Original equation}$$

$$-20 = 4(0 - 5) \quad \text{Replace } y \text{ with } 0.$$

$$-20 = 4(-5)$$

$$-20 = -20 \quad \text{True}$$

0 is the correct solution (not –20).

Objective 1   Solve equations with several steps.

For extra help, see Example 1 on page 674 of your text and Section Lecture video for Section 9.7 and Exercise Solutions Clip 3.

Solve each equation. Check each solution.

1. $13 = 2y - 9$ 1. _____

2. $6k + 4 = 16$ 2. _____

3. $5p - 4.2 = -17.7$ 3. _____

Objective 2   Use the distributive property.

For extra help, see Example 2 on page 675 of your text and Section Lecture video for Section 9.7 and Exercise Solutions Clip 15 and 21.

Use the distributive property to simplify.

4. $-5(2+a)$ 4. _____

5. $-4(8-x)$ 5. _____

6. $7(k-5)$ 6. _____

Objective 3   Combine like terms.

For extra help, see Example 3 on page 675 of your text and Section Lecture video for Section 9.7.

Combine like terms.

7. $9z-4z$ 7. _____

8. $-3.2x+1.3x$ 8. _____

9. $5a-2.7a$ 9._____

Objective 4   Solve more difficult equations.

For extra help, see Examples 4–5 on pages 676–677 of your text and Section Lecture video for Section 9.7 and Exercise Solutions Clip 51 and 61.

Solve each equation. Check each solution.

10. $6y-13y=-14$ 10._____

11. $-3.7z - 0.5 = -5.2z + 4$

11. _____

12. $-12 = 0.2(5 - x)$

12. _____

Chapter 9 BASIC ALGEBRA

9.8   Using Equations to Solve Application Problems

Learning Objectives
1   Translate word phrases into expressions with variables.
2   Translate sentences into equations.
3   Solve application problems.

Key Terms

Use the vocabulary terms listed below to complete each statement in exercises 1–5.

indicator words	**sum**	**difference**	**product**	**quotient**
increased by	**less than**	**double**	**per**	

1.   Words in a problem that indicate the necessary operations are

_____.

2.   _____ and _____ are indicator words for
addition.

3.   _____ and _____ are indicator words for
multiplication.

4.   _____ and _____ are indicator words for
division.

5.   _____ and _____ are indicator words for
subtraction.

Guided Examples

Review these examples for Objective 1:

1.   Write each word phrase in symbols, using x as the variable.

    a. Words: A number plus 11

    Algebraic Expression: $x + 11$ or $11 + x$

    b. Words: The sum of 5 and a number

    Algebraic Expression: $5 + x$ or $x + 5$

    c. Words: 9 more than a number

    Algebraic Expression: $x + 9$ or $9 + x$

Now Try:

1.   Write each word phrase in symbols, using x as the variable.

    a. 16 less than a number

    _____

    b. The number plus 19

    _____

    c. 20 minus a number

    _____

d. Words: –49 added to a number

Algebraic Expression: $-49 + x$ or $x + (-49)$

e. Words: A number increased by 10

Algebraic Expression: $x + 10$ or $10 + x$

f. Words: 18 less than a number

Algebraic Expression: $x - 18$

g. Words: A number subtracted from 17

Algebraic Expression: $17 - x$

h. Words: A number decreased by 16

Algebraic Expression: $x - 16$

i. Words: 37 minus a number

Algebraic Expression: $37 - x$

2. Write each word phrase in symbols, using x as the variable.

   a. Words: 6 times a number

   Algebraic Expression: $6x$

   b. Words: The product of 25 and a number

   Algebraic Expression: $25x$

   c. Words: Triple a number

   Algebraic Expression: $3x$

   d. Words: Three times a number

   Algebraic Expression: $3x$

   e. Words: The quotient of 7 and a number

   Algebraic Expression: $\dfrac{7}{x}$

d. 12 more than a number

_____

e. A number increased by 13

_____

f. A number decreased by 27

_____

g. –54 added to a number

_____

h. The sum of 99 and a number

_____

i. A number subtracted from 38

_____

2. Write each word phrase in symbols, using x as the variable.

   a. The product of –7 and a number

   _____

   b. The product of a number and 5

   _____

   c. Five times a number

   _____

   d. The quotient of 16 and a number

   _____

   e. A number divided by 24

   _____

f. Words: A number divided by 14

Algebraic Expression: $\dfrac{x}{14}$

g. Words: One-fourth of a number

Algebraic Expression: $\dfrac{1}{4}x$ or $\dfrac{x}{4}$

h. Words: The result is

Algebraic Expression: =

f. The quotient of –11 and a number

_____

g. One-fifth of a number

_____

h. A number divided by 6

_____

Review this example for Objective 2:

3. If twice a number is decreased by 3, the result is –17. Find the number.

Let x represent the unknown number. Use the information in the problem to write an equation.

$$\underbrace{\text{Twice a number}}\ \underbrace{\text{decreased by}}\ 3\ \text{is}\ -17$$
$$\downarrow\qquad\qquad\downarrow\qquad\ \downarrow\ \downarrow\ \downarrow$$
$$2x\qquad\qquad\ -\qquad 3\ =\ -17$$

Next, solve the equation.

$$2x - 3 + 3 = -17 + 3 \quad \text{Add 3 to both sides.}$$
$$2x = -14$$

$$\dfrac{\overset{1}{\cancel{2}}\,x}{\underset{1}{\cancel{2}}} = \dfrac{-14}{2}$$

$$x = -7$$

The unknown number is –7.
To check the solution, go back to the words of the original problem.

$$\text{If}\ \underbrace{\text{twice}\ \underbrace{\text{a number}}}\ \underbrace{\text{decreased by}}\ 3\ \text{the result is}-17$$
$$\downarrow\qquad\downarrow\qquad\qquad\downarrow\qquad\downarrow\qquad\downarrow\qquad\qquad\downarrow$$
$$2\quad\ (-7)\qquad\ -\qquad 3\quad =\qquad -17$$

Does $2(-7) - 3$ really equal –17? Yes,
$-14 - 3 = -17$. So, –7 is the correct solution.

Now Try:

3. If the product of some number and 2 is increased by 18, the result is four times the number. Find the number.

_____

Review these examples for Objective 3:

4. A rental car costs $32 per day plus $0.20 per mile. The bill for a one-day rental was $82. How many miles was the car driven?

    Step 1 Read. The problem asks for the number of miles the car was driven.

    Step 2 Assign a variable. There is only one unknown: the number of miles the car was driven. Let x represent the number of miles the car was driven.

    Step 3 Write an equation.

    $$\underbrace{\text{Fixed cost}} \;\; \text{and} \;\; \$0.20 \;\; \underbrace{\text{per mile}} \;\; \text{is} \;\; \$82$$
    $$\$32 \;\;\;\; + \;\;\;\; \$0.20 \;\;\;\; \cdot \; x \;\;\;\; = \;\; \$82$$

    Step 4 Solve.
    $$32 + 0.20x = 82$$
    $$32 + 0.20x - 32 = 82 - 32$$
    $$0.20x = 50$$
    $$\frac{\overset{1}{\cancel{0.20}} \cdot x}{\underset{1}{\cancel{0.20}}} = \frac{50}{0.20}$$
    $$x = 250$$

    Step 5 State the answer. The car was driven 250 miles.

    Step 6 Check
    $$\underbrace{\text{Fixed cost}} \;\; \text{and} \;\; \$0.20 \;\; \underbrace{\text{per mile}} \;\; \text{is} \;\; \$82$$
    $$\$32 \;\;\;\; + \;\; \$0.20 \;\; \cdot \; 250 \;\; = \;\; \$82$$
    $$\$32 \;\; + \;\; \$50 \;\;\;\;\;\;\;\;\; = \$82$$
    $$\$82 \;\;\;\;\;\;\;\;\;\;\;\;\;\;\; = \$82 \; \text{True}$$

    So 250 miles is the correct solution because it "works" in the original problem.

Now Try:

4. The larger of two numbers is twice the smaller number. The sum of the numbers is 36. Find the smaller number.

    _____

5. If half a number is added to twice the number, the result is 55. Find the number.

Step 1 Read. The problem asks for the number.

Step 2 Assign a variable. There is only one unknown: the number.
Let x represent the number.

Step 3 Write an equation.

Half the number	added to	twice the number	is	55
$\frac{1}{2}x$	$+$	$2x$	$=$	55

Step 4 Solve.

$$\frac{1}{2}x + 2x = 55$$

$$\left(\frac{1}{2} + 2\right)x = 55$$

$$\frac{5}{2}x = 55$$

$$\frac{\cancel{2}}{\cancel{5}} \cdot \frac{\cancel{5}}{\cancel{2}}x = \frac{55}{1} \cdot \frac{2}{5}$$

$$x = 22$$

Step 5 State the answer. The number is 22.

Step 6 Check.

Half of 22	added to	twice 22	is	55
$\frac{1}{2}(22)$	$+$	$2(22)$	$=$	55
11	$+$	44	$=55$	
	55		$=55$	

So 22 is the correct solution because it "works" in the original problem.

6. Lisa and Michael were opposing candidates for city council. Lisa won, with 73 more votes than Michael. The total number of votes received by both candidates was 567. Find the number of votes received by Michael.

Step 1 Read. The problem asks for the number of votes received by Michael.

5. If ten times a number is subtracted from six times the number, the result is 12. Find the number.

_____

6. Mrs. Wong's class read eighteen less than twice as many books as Mr. Lee's class read. If Mrs. Wong's class read 40 books, how many books did Mr. Lee's class read?

_____

Step 2 Assign a variable. There are two unknowns: the number of votes received by Michael and the number of votes received by Lisa.

Let x be the number of votes received by Michael.

Since Lisa received 73 more votes than Michael, the number of votes she received is $x + 73$, that is, votes for Michael (x) plus 73.

Step 3 Write an equation.

Votes for Michael	plus	Votes for Lisa	is	567
↓	↓	↓	↓	↓
x	$+$	$x+73$	$=$	567

Step 4 Solve.

$$x + x + 73 = 567$$
$$2x + 73 = 567$$
$$2x + 73 - 73 = 567 - 73$$
$$2x = 494$$
$$\frac{\overset{1}{\cancel{2}}x}{\underset{1}{\cancel{2}}} = \frac{494}{2}$$
$$x = 247$$

Step 5 State the answer.
Michael received x votes, so there were 247 votes for Michael.
Lisa received $x + 73$ votes, so Lisa received $247 + 73 = 320$ votes.

Step 6 Check. Use the words in the original problem. "Lisa won, with 73 more votes than Michael." Lisa's 320 votes is 73 more than Michael's 247 votes.
"The total number of votes received by both candidates was 567."
Michael's 247 + Lisa's 320 = 567 votes, so that checks.

7. The length of a rectangle is 3 in. more than the width of the rectangle. If the perimeter of the rectangle is 26 in., find the length and the width of the rectangle.

   Step 1 Read. The problem asks for the length

7. The length of a rectangle is 27 centimeters, while the perimeter is 70 centimeters. Find the width of the rectangle.

---

and width of the rectangle.

Step 2 Assign a variable. There are two unknowns, length and width. You know the least about the width, so let x represent the width. Since the length is 3 in. more than the width, the length is $x + 3$.

Step 3 Write an equation. Use the formula for perimeter of a rectangle, $P = 2l + 2w$.

$$P = 2 \cdot l + 2 \cdot w$$

Step 4 Solve.

$$26 = 2 \cdot (x + 3) + 2 \cdot x$$
$$26 = 2x + 6 + 2x$$
$$26 = 4x + 6$$
$$26 - 6 = 4x + 6 - 6$$
$$20 = 4x$$
$$\frac{20}{4} = \frac{\overset{1}{\cancel{4}} \cdot x}{\underset{1}{\cancel{4}}}$$
$$5 = x$$

Step 5 State the answer. The width is x, so the width is 5 in.

The length is $x + 3$, so the length is $5 + 3$ or 8 in. The width is 5 in. and the length is 8 in.

Step 6 Check. Use the words of the original problem. It says that the length is 3 in. more than the width. 8 in. is 3 more than 5 in., so that part checks.

The original problem also says the perimeter is 26 in. Use 5 in. and 8 in. to find the perimeter.

$$P = 2 \cdot 8 \text{ in.} + 2 \cdot 5 \text{ in.}$$

$$P = 16 \text{ in.} + 10 \text{ in.} = 26 \text{ in.} \leftarrow \text{Checks}$$

Objective 1   Translate word phrases into expressions with variables.

For extra help, see Examples 1–2 on page 682 of your text and Section Lecture video for Section 9.8 and Exercise Solutions Clip 3, 5, 17, and 19.

Write each word phrase in symbols, using ***x*** *as the variable.*

1.   The sum of 9 and a number

1. _____

2. 5 subtracted from a number   2. _____

3. The product of a number and 3   3. _____

Objective 2   Translate sentences into equations.

For extra help, see Example 3 on page 683 of your text and Section Lecture video for Section 9.8 and Exercise Solutions Clip 25.

Translate each sentence into an equation and solve it. Check your solution by going back to the words in the original problem.

4. If –5 times a number is added to 4, the result is –11.   **4.**
Find the number.

Equation _____

Solution _____

5. If twice a number is subtracted from 45, the result   **5.**
is 35. Find the number.

Equation _____

Solution _____

6. When two times a number is subtracted from 8, the   **6.**
result is 20 plus the number. Find the number.

Equation _____

Solution _____

Objective 3   Solve application problems.

For extra help, see Examples 4–7 on pages 684–687 of your text and Section Lecture video for Section 9.7 and Exercise Solutions Clip 39 and 51.

Solve each application problem using the six problem-solving steps listed in the text.

7. The price of a DVD is $3.00 less than twice the cost   7. _____
of a book. If the DVD costs $25.00, how much does
the book cost?

8. A board is 91 centimeters long. It is to be cut into
    two pieces, with one piece 15 centimeters longer
    than the other. Find the length of the shorter piece.

8. _____

9. When the difference between a number and 4 is
    multiplied by −3, the result is two more than −5
    times the number. Find the number.

9. _____

Chapter 10 STATISTICS

10.1 Circle Graphs

> **Learning Objectives**
> 1   Read and understand a circle graph.
> 2   Use a circle graph.
> 3   Draw a circle graph.

Key Terms

Use the vocabulary terms listed below to complete each statement in exercises 1–2.

     circle graph   **protractor**

1.   A _____ shows how a total amount is divided into parts or sectors.

2.   A _____ is a device used to measure the number of degrees in angles or parts of a circle.

Guided Examples

The circle graph shows the number of students enrolled in certain majors at a college. The entire circle represents 11,600 students.

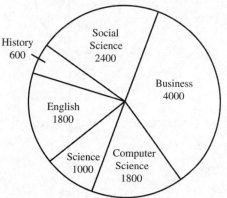

Review these examples for Objective 2:

1. Find the ratio of social science majors to the total number of students. Write the ratio as a fraction in lowest terms.

   The circle graph shows 2400 social science majors of the 11,600 total students. The ratio of social science students to the total number of students is shown below.

   $$\frac{2400 \text{ students (social science)}}{11{,}600 \text{ students (total)}}$$

   $$= \frac{2400 \text{ students}}{11{,}600 \text{ students}} = \frac{2400 \div 400}{11{,}600 \div 400} = \frac{6}{29}$$

Now Try:

1. Find the ratio of history majors to the total number of students. Write the ratio as a fraction in lowest terms.

   _____

2. Use the circle graph to find the ratio of history majors to business majors. Write the ratio as a fraction in lowest terms.

   The circle graph shows 600 history majors and 4000 business majors. The ratio of history majors to business majors is shown below.

   $$\frac{600 \text{ students (history)}}{4000 \text{ students (business)}}$$

   $$= \frac{600 \text{ students}}{4000 \text{ students}} = \frac{600 \div 200}{4000 \div 200} = \frac{3}{20}$$

2. Use the circle graph to find the ratio of science majors to computer science majors. Write the ratio as a fraction in lowest terms.

   _____

The circle graph shows the expenses involved in keeping a sales force on the road. Each expense item is expressed as a percent of the total sales force cost of $950,000.

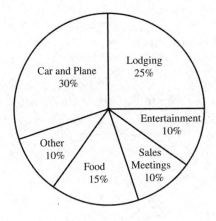

3. Use the circle graph above on expenses to find the amount spent on lodging.

   Recall the percent equation.
       part = percent · whole
   The total expenses is $950,000, so the whole is $950,000. The percent is 25%, or as a decimal, 0.25. Find the part.
       part = percent · whole

   $$x = (0.25)(950,000)$$

   $$x = 237,500$$

   The amount spent on lodging was $237,500.

3. Use the circle graph above on expenses to find the amount spent on food.

   _____

Review this example for Objective 3:

4. A family recorded its expenses for a year, with the following results: 40% for Housing, 20% for Food, 14% for Automobile, 8% for Clothing, 6% for Medical, 8% for Savings, and 4% for Other.

    a. Find the number of degrees in a circle graph for each type of expenses.

    Recall that a complete circle has 360°. Because "Housing" makes up 40% of the expenses, the number of degrees needed for the "Housing" sector of the circle graph is 40% of 360°.

    Housing

    $$(360°)(40\%) = (360°)(0.40) = 144°$$

    Food

    $$(360°)(20\%) = (360°)(0.20) = 72°$$

    Automobile

    $$(360°)(14\%) = (360°)(0.14) = 50.4°$$

    Clothing

    $$(360°)(8\%) = (360°)(0.08) = 28.8°$$

    Medical

    $$(360°)(6\%) = (360°)(0.06) = 21.6°$$

    Savings

    $$(360°)(8\%) = (360°)(0.08) = 28.8°$$

    Other

    $$(360°)(4\%) = (360°)(0.04) = 14.4°$$

    b. Draw a circle graph showing this information.

    Use a protractor to make the circle graph.

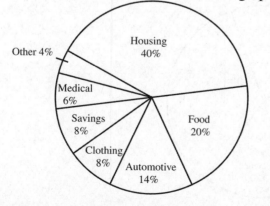

Now Try:

4. A book publisher had 30% of its sales in mysteries, 15% in biographies, 10% in cookbooks, 25% in romance novels, 15% in science, and the rest in business books.

    a. Find the number of degrees in a circle graph for each type of book.

    mysteries _____

    biographies _____

    cookbooks _____

    romance _____

    science _____

    business _____

    b. Draw a circle graph showing this information.

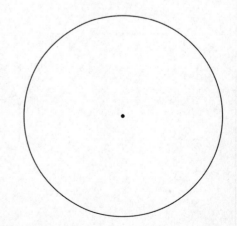

Name: Date:
Instructor: Section:

Objective 1   Read and understand a circle graph.

For extra help, see page 708 of your text and Section Lecture video for Section 10.1.

The circle graph shows the cost of remodeling a kitchen. Use the graph to answer exercises 1–3.

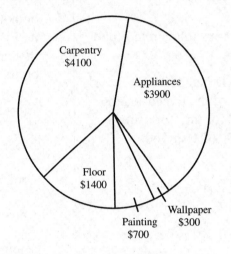

1.   Find the total cost of remodeling the kitchen.   1. _____

2.   What is the largest single expense in remodeling the   2. _____
      kitchen?

3.   How much less does the wallpaper cost than   3. _____
      painting?

Objective 2   Use a circle graph.

For extra help, see Examples 1–3 on pages 708–709 of your text and Section Lecture video for Section 10.1 and Exercise Solutions Clip 17, 19, and 21.

The circle graph shows the number of students enrolled in certain majors at a college. Use the graph to answer exercises 4–5. The entire circle represents 11,600 students.

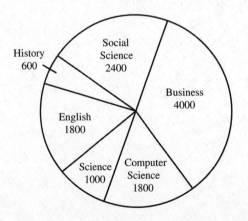

4. Find the ratio of the number of business majors to the total number of students.

4. _____

5. Find the ratio of the number of science majors to the number of English majors.

5. _____

The circle graph shows the expenses involved in keeping a sales force on the road. Each expense item is expressed as a percent of the total sales force cost of $950,000. Find the number of dollars of expense for the category.

6. Car and plane

6. _____

Objective 3   Draw a circle graph.

For extra help, see Example 4 on pages 710–711 of your text and Section Lecture video for Section 10.1 and Exercise Solutions Clip 39.

Use the given information to draw a circle graph.

Jensen Manufacturing Company has its annual sales divided into five categories as follows. The total sales for a year is $400,000.

Item	Annual Sales
Parts	$20,000
Hand tools	80,000
Bench tools	100,000
Brass fittings	140,000
Cabinet hardware	60,000

7. Find the percent of the total sales for each item.

7. parts _____

hand tools _____

bench tools _____

brass fittings _____

hardware _____

8. Find the number of degrees in a circle graph for each item.

8. parts _____

hand tools _____

bench tools _____

brass fittings _____

hardware _____

9. Make a circle graph showing this information.

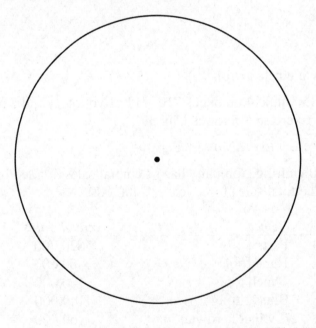

Chapter 10 STATISTICS

10.2   Bar Graphs and Line Graphs

Learning Objectives
1   Read and understand a bar graph.
2   Read and understand a double-bar graph.
3   Read and understand a line graph.
4   Read and understand a comparison line graph.

Key Terms

Use the vocabulary terms listed below to complete each statement in exercises 1–4.

bar graph   double-bar graph   line graph   comparison line graph

1.   A _____ uses dots connected by lines to show trends.

2.   A _____ compares two sets of data by showing two sets of bars.

3.   A _____ uses bars of various heights or lengths to show quantity or frequency.

4.   A _____ shows how two sets of data relate to each other by showing a line graph for each item.

Guided Examples

The bar graph shows the enrollment for the fall semester at a small college from 2002 to 2006.

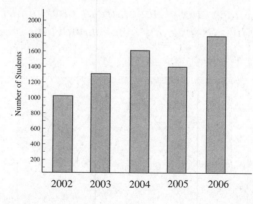

Review this example for Objective 1:

1.   What was the fall enrollment for 2003?

     The bar for 2003 rises to 1300. So the enrollment in 2003 was 1300 students.

Now Try:

1.   What was the fall enrollment for 2006?

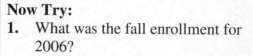

Name: _____   Date:

Instructor: _____   Section:

The double-bar graph shows the enrollment by gender in each class at a small college.

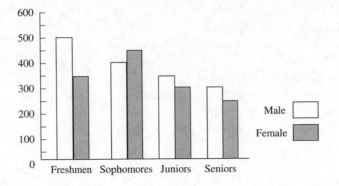

Review these examples for Objective 2:

2. Use the double-bar graph to find the following.

   a. The number of male sophomores enrolled

There are two bars for the sophomore class. The color code to the right tells you that white bars represent male enrollments. So the white bar on the left for Sophomores represents the number of male sophomores. It rises to 400.
So the sophomore enrollment of males is 400.

   b. The number of female juniors enrolled

The gray column for juniors rises to 300.
So the junior enrollment of females is 300.

Now Try:

2. Use the double-bar above to find the following.

   a. The number of female seniors enrolled

   _____

   b. The number of male freshmen enrolled

   _____

The line graph gives the value of one share of stock of Microchip Computer Corporation on the first trading day of the month for six consecutive months.

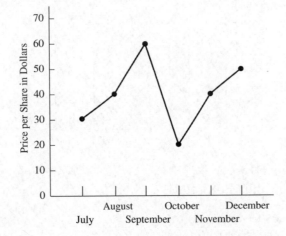

Name: Date:
Instructor: Section:

Review these examples for Objective 3:

3. Use the line graph on the previous page to find the following.

   a. In which month was the value of the stock the lowest?

   The lowest point on the graph is the dot directly over October, so the lowest value of the stock occurred in October.

   b. Find the value of one share of stock on the first trading day in August.

   Use a ruler or straightedge to line up the August dot with the numbers along the left edge of the graph. The August dot is directly across from 40. So in August, the price per share was $40.

Now Try:

3. Use the line graph on the previous page to find the following.
   a. Find the value of the stock in September.

   _____

   b. In which month was the value of the stock $50?

   _____

The comparison line graph shows annual sales for two different stores from 2002 to 2006.

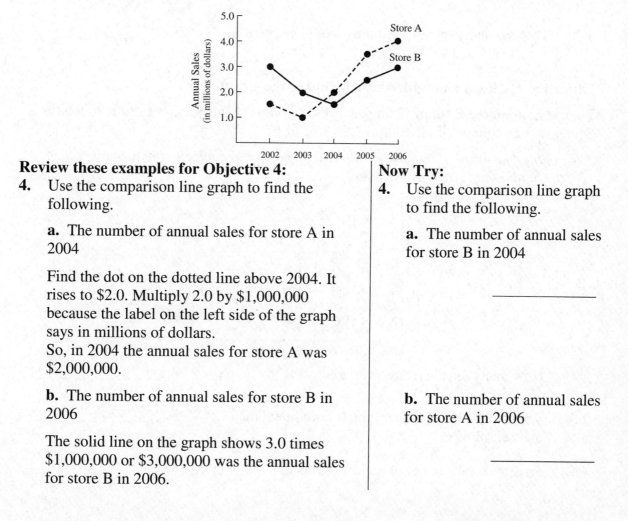

Review these examples for Objective 4:

4. Use the comparison line graph to find the following.

   a. The number of annual sales for store A in 2004

   Find the dot on the dotted line above 2004. It rises to $2.0. Multiply 2.0 by $1,000,000 because the label on the left side of the graph says in millions of dollars.
   So, in 2004 the annual sales for store A was $2,000,000.

   b. The number of annual sales for store B in 2006

   The solid line on the graph shows 3.0 times $1,000,000 or $3,000,000 was the annual sales for store B in 2006.

Now Try:

4. Use the comparison line graph to find the following.

   a. The number of annual sales for store B in 2004

   _____

   b. The number of annual sales for store A in 2006

   _____

Name: Date:
Instructor: Section:

Objective 1   Read and understand a bar graph.

For extra help, see Example 1 on page 718 of your text and Section Lecture video for Section 10.2 and Exercise Solutions Clip 3, 5, and 7.

The bar graph shows the enrollment for the fall semester at a small college from 2002 to 2006. Use this graph for problems 1−3.

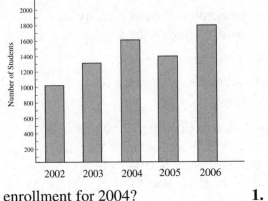

1.   What was the fall enrollment for 2004?   1. _____

2.   What year had the greatest enrollment?   2. _____

3.   By how many students did the enrollment increase   3. _____
     from 2005 to 2006?

Objective 2   Read and understand a double-bar graph.

For extra help, see Example 2 on page 719 of your text and Section Lecture video for Section 10.2 and Exercise Solutions Clip 15 and 17.

The double-bar graph shows the enrollment by gender in each class at a small college. Use the double-bar graph for Problems 4−6.

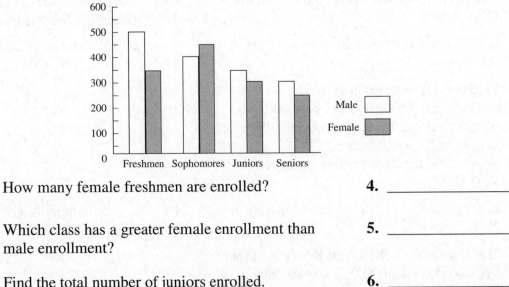

4.   How many female freshmen are enrolled?   4. _____

5.   Which class has a greater female enrollment than 5. _____
     male enrollment?

6.   Find the total number of juniors enrolled.   6. _____

Copyright © 2014 Pearson Education, Inc.

Name: Date:
Instructor: Section:

Objective 3   Read and understand a line graph.

For extra help, see Example 3 on page 719 of your text and Section Lecture video for
Section 10.2 and Exercise Solutions Clip 21 and 23.

*The line graph gives the value of one share of stock of Microchip Computer Corporation
on the first trading day of the month for six consecutive months. Use the line graph for
Problems 7–9.*

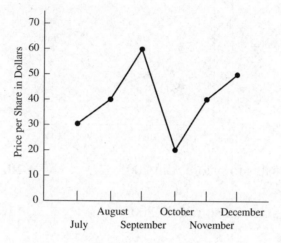

7.   In which month was the value of the stock highest?   7. _____

8.   Find the value of one share on the first trading day   8. _____
      October.

9.   By how much did the value of one share increase   9. _____
      from July to September?

Objective 4   Read and understand a comparison line graph.

For extra help, see Example 4 on page 720 of your text and Section Lecture video for Section 10.2.

The comparison line graph shows annual sales for two different stores from 2002 to 2006. Use the graph to solve Problems 10–12.

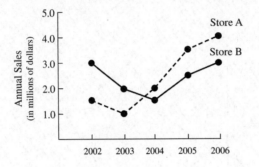

10.   Find the annual sales for store A in 2005.   10. _____

11.   Find the annual sales for store B in 2003.   11. _____

12.   In which years did the sales of store *A* exceed the   12. _____
       sales of store *B*?

Chapter 10 STATISTICS

10.3   Frequency Distributions and Histograms

Learning Objectives
1   Understand a frequency distribution.
2   Arrange data in class intervals.
3   Read and understand a histogram.

Key Terms

Use the vocabulary terms listed below to complete each statement in exercises 1–2.

frequency distribution   histogram

1.   A bar graph in which the width of each bar represents a range of number and the height represents the quantity or frequency of items that fall within the interval is called a _____.

2.   A table that includes a column showing each possible number in the data collected is called a _____.

Guided Examples

Review this example for Objective 1:

1.   A teacher has kept track of her online quiz scores for 35 students. The score for each student is given below.

75	85	90	60	85	85	60
75	50	80	85	95	92	95
75	50	90	75	90	62	95
80	80	65	65	68	60	85
75	65	90	62	85	60	92

Construct a frequency distribution table.

Using the data above, construct a table that shows each possible score. Then go through the original data and place a tally mark in the tally column next to each corresponding value. Total the tally marks and place the totals in the third column.

Now Try:

1.   The following list of numbers represents IQ scores of 18 students.

98	121	112
102	102	112
106	102	92
112	102	106
106	92	92
102	98	112

Use these scores to construct a frequency distribution table.

Scores	Tally	Frequency
50	||	2
60	||||	4
62	||	2
65	|||	3
68	|	1
75	Ж	5
80	|||	3
85	Ж |	6
90	||||	4
92	||	2
95	|||	3

The online quiz score data can be combined into groups, forming class intervals.

Interval	Frequency
50 – 59	2
60 – 69	10
70 – 79	5
80 – 89	9
90 – 99	9

Review these examples for Objective 2:

2. Use the grouped data to answer the following questions.

   a. How many students earned scores below 60?

   The first interval in the grouped data table (50–59) is the scores where students earned below 60. Therefore, 2 students earned less than 60.

   b. How many students earned scores of 70 or more?

   The last three intervals are scores of 70 or more.
   $5 + 9 + 9 = 23$ students

Now Try:

2. Use the grouped data to answer the following questions.

   a. How many students earned scores less than 80?

   _____

   b. How many students scored 80 or more?

   _____

The IQ scores of 18 students were recorded and a histogram was constructed.

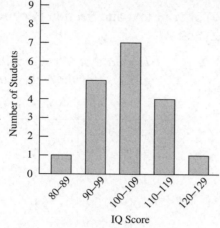

IQ Score

Review this example for Objective 3:

3. Use the histogram to find the number of students who had IQ score of less than 100.

   The histogram shows that a 80–89 score was achieved by 1 student and a 90–99 score was achieved by 5 students. So, the number of students who had IQ scores less than 100 is
   1 + 5 = 6 students

Now Try:

3. Use the histogram to find the number of students who had IQ score of greater than 109.

   _____

Objective 1   Understand a frequency distribution.

For extra help, see Example 1 on page 729 of your text and Section Lecture video for Section 10.3 and Exercise Solutions Clip 17 and 21.

The following scores were earned by students on an algebra exam. Use the data to find the tally and the frequency for the given score in problems 1–3.

84	90	83	72	84	93	83	90	83
90	72	64	90	83	72	83	83	64

1. 72

1. tally_____

    frequency _____

2. 83

2. tally_____

    frequency _____

3. 93

3. tally_____

    frequency _____

Name: _____   Date: _____
Instructor: _____   Section: _____

Objective 2   Arrange data in class intervals.

For extra help, see Example 2 on page 730 of your text and Section Lecture video for Section 10.2 and Exercise Solutions Clip 23 and 27.

The following list of numbers represents systolic blood pressure of 21 patients.

120	98	180
128	143	98
105	136	115
190	118	105
180	112	160
110	138	122
98	175	118

The data can be combined into groups, forming class intervals.

Interval	Frequency
90 – 109	5
110 – 129	8
130 – 149	3
150 – 169	1
170 – 189	3
190 – 209	1

4. How many patients had systolic blood pressure less than 110?

4. _____

5. How many patients had systolic blood pressure greater than 150?

5. _____

6. How many patients had systolic blood pressure less than 150?

6. _____

Name: Date:
Instructor: Section:

Objective 3   Read and understand a histogram.

For extra help, see Example 3 on page 730 of your text and Section Lecture video for Section 10.3.

The systolic blood pressure was recorded for 21 patients and a histogram was constructed. Use the histogram to solve Problems 7–9.

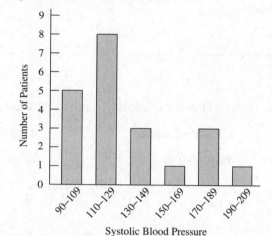

7.   Use the histogram to find the number of patients   7. _____
     who had systolic blood pressure of less than 130.

8.   Use the histogram to find the number of patients   8. _____
     who had systolic blood pressure of greater than 129.

9.   Use the histogram to find the number of patients   9. _____
     who had systolic blood pressure of greater than 170.

Chapter 10 STATISTICS

10.4   Mean, Median, and Mode

Learning Objectives
1   Find the mean of a list of numbers.
2   Find a weighted mean.
3   Find the median.
4   Find the mode.

Key Terms

Use the vocabulary terms listed below to complete each statement in exercises 1–7.

mean	weighted mean	median	mode
bimodal	**dispersion**	**range**	

1.   The _____ is the variation or spread of the numbers around the mean.

2.   The _____ is the value that occurs most often in a group of values.

3.   A mean calculated so that each value is multiplied by its frequency is called a _____.

4.   The sum of all the values in a data set divided by the number of values in the data set is called the _____.

5.   The middle number in a group of values that are listed from least to greatest is called the _____.

6.   When two values in a data set occur the same number of times, the data set is called _____.

7.   The difference between the greatest value and the least value in a set of numbers is called the _____.

Guided Examples

Review these examples for Objective 1:

1.   Find the mean of 125, 234, 155, 275, and 141.

Use the formula for finding the mean. Add up all the values and then divide by the number of values.

$$\text{mean} = \frac{125 + 234 + 155 + 275 + 141}{5}$$

Now Try:

1.   Find the mean of 257, 261, 269, 274, 268, 280, and 295.

_____

$$mean = \frac{930}{5}$$

$$mean = 186$$

The mean is 186.

2. Find the mean of 31, 37, 44, 51, 52, 74, 69, and 83.

   Find the mean (rounded to the nearest tenth).

   $$mean = \frac{31+37+44+51+52+74+69+83}{8}$$

   $$mean = \frac{441}{8}$$

   $$mean \approx 55.1$$

   The mean is 55.1.

2. Find the mean of 40.1, 32.8, 82.5, 51.2, 88.3, 31.7, 43.7, and 51.2.

   _____

Review these examples for Objective 2:

3. Use the following table to find the weighted mean.

Value	Frequency
13	4
12	2
19	5
15	3
21	1
27	5

To find the mean, multiply the value by the frequency. Then add the products. Next, add the numbers in the frequency column to find the total number of values.

Value	Frequency	Product
13	4	$(13 \cdot 4) = 52$
12	2	$(12 \cdot 2) = 24$
19	5	$(19 \cdot 5) = 95$
15	3	$(15 \cdot 3) = 45$
21	1	$(21 \cdot 1) = 21$
27	5	$(27 \cdot 5) = 135$
Totals	20	372

Finally, divide the totals. Round to the nearest tenth.

$$mean = \frac{372}{20} \approx 18.6$$

The mean is 18.6.

Now Try:

3. Use the following table to find the weighted mean.

Value	Frequency
35	1
36	2
39	5
40	4
42	3
43	5

_____

4. Find the grade point average for this student.
Assume A = 4, B = 3, C = 2, D = 1, F = 0.

Units	Grade
3	C
3	A
4	B
5	B
2	A

Multiply the credits and grades. Find the total.

Units	Grade	Units · Grade
3	C $(= 2)$	$3 \cdot 2 = 6$
3	A $(= 4)$	$3 \cdot 4 = 12$
4	B $(= 3)$	$4 \cdot 3 = 12$
5	B $(= 3)$	$5 \cdot 3 = 15$
2	A $(= 4)$	$2 \cdot 4 = 8$
Totals 17		53

It is common to round grade point averages to the nearest hundredth.

$$\text{GPA} = \frac{53}{17} \approx 3.12$$

4. Find the grade point average for this student. Assume A = 4, B = 3, C = 2, D = 1, F = 0.

Units	Grade
3	A
4	B
2	C
5	C
2	D

Review these examples for Objective 3:

5. Find the median for the list of values.

21, 32, 27, 23, 25, 29, 22

First arrange the numbers in numerical order from least to greatest.

21, 22, 23, 25, 27, 29, 32

Next, find the middle number in the list.

21, 22, 23, 25, 27, 29, 32

Three are below. ↓ Three are above.

Middle number

The median value is 25.

6. Find the median for the list of values.

389, 464, 521, 610, 654, 672, 682, 712

First arrange the numbers in numerical order from least to greatest. Then find the middle two numbers.

389, 464, 521, 610, 654, 672, 682, 712

Middle two numbers

The median value is the mean of the two middle

Now Try:

5. Find the median for the list of values.

18, 12, 11, 19, 26

_____

6. Find the median for the list of values.

0.02, 0.04, 0.12, 0.08

_____

numbers.

$$\text{median} = \frac{610 + 654}{2} = \frac{464}{2} = 632$$

The median value is 632.

Review these examples for Objective 4:	Now Try:
7. Find the mode for each list of numbers.	**7.** Find the mode for each list of numbers.
a. 37, 52, 41, 27, 41, 96	**a.** 5, 3, 6, 3, 7, 8, 5, 3, 4
The number 41 occurs more often than any other number; therefore, 41 is the mode.	_____
b. 964, 987, 973, 987, 921, 921, 975	**b.** 16, 13, 21, 16, 18, 11, 13, 15, 14
Because both 987 and 921 occur twice, each is a mode.	_____
c. $10.71, $11.67, $12.39, $21.54, $13.98, $14.66	**c.** 1.5, 1.2, 1.3, 1.9, 1.8, 1.4, 1.7, 1.0
No number occurs more than once. This list has no mode.	_____

Objective 1   Find the mean of a list of numbers.

For extra help, see Examples 1–2 on page 735 of your text and Section Lecture video for Section 10.4 and Exercise Solutions Clip 9 and 11.

Find the mean for each list of numbers. Round to the nearest tenth, if necessary.

1.   39, 50, 59, 61, 69, 73, 51, 80   1. _____

2.   62.7, 59.6, 71.2, 65.8, 63.1   2. _____

3.   216, 245, 268, 268, 280, 291, 304, 313   3. _____

Objective 2   Find a weighted mean.

For extra help, see Examples 3–4 on pages 736–737 of your text and Section Lecture video for Section 10.4.

Find the weighted mean for each list of numbers. Round to the nearest tenth, if necessary.

4.   4. _____

Value	Frequency
17	4
12	5
15	3
19	1

5.

Value	Frequency
1	2
2	3
4	5
5	7
6	4
7	2
8	1
9	1

5. _____

Find the grade point average for this student. Assume A = 4, B = 3, C = 2, D = 1, F = 0.

6.

Units	Grade
5	B
4	C
3	B
2	C
2	C

6. _____

Objective 3   Find a median.

For extra help, see Examples 5–6 on pages 737–738 of your text and Section Lecture video for Section 10.4.

Find the median for each list of numbers.

7.   200, 215, 226, 238, 250, 283   7. _____

8.   43, 69, 108, 32, 51, 49, 83, 57, 64   8. _____

9.   200, 195, 302, 284, 256, 237, 239, 240   9. _____

Objective 4   Find the mode.

For extra help, see Example 7 on page 738 of your text and Section Lecture video for Section 10.4 and Exercise Solutions Clip 25.

Find the mode for each list of numbers.

10.   4, 9, 3, 4, 7, 3, 2, 3, 9   10. _____

11.   37, 24, 35, 35, 24, 38, 39, 28, 27, 39   11. _____

12.   172.6, 199.7, 182.4, 167.1, 172.6, 183.4, 187.6   12. _____

Chapter 1 WHOLE NUMBERS

1.1   Reading and Writing Whole Numbers
Key Terms
1. table
2. whole numbers
3. place value

Now Try
1. 13 is a whole number
2a. hundreds
2b. ones
2c. tens
3. 3 ten-thousands, 6 thousands, 5 hundreds, 4 tens, 9 ones
4. 80 billions, 429 millions, 321 thousands, 635 ones
5a. twelve
5b. sixty-three
5c. nine hundred fourteen
5d. two hundred eight
6a. three hundred forty-eight thousand, six hundred twenty-nine
6b. two thousand, eight hundred, fifty
6c. eight hundred-million, eighteen thousand, six
6d. fourteen billion, twenty-one million, sixty thousand
7a. 7034
7b. 563,916
7c. 2,000,974
8a. three hundred forty-two
8b. one hundred ninety-five

Objective 1
1. not a whole number
3. whole number

Objective 2
5. 2; 0

Objective 3
7. eight hundred thirty-four thousand, seven hundred sixty-eight
9. 972,430

Objective 4
11. three hundred fifteen

Answers

1.2   Adding Whole Numbers

Key Terms

1. commutative property of addition

2. addends

3. addition

4. associative property of addition

5. regrouping

6. perimeter

7. sum

Now Try

1a. $7 + 2 = 9$ and $2 + 7 = 9$

1b. $5 + 9 = 14$ and $9 + 5 = 14$

1c. $3 + 3 = 6$

2. 29

3. 788

4. 65

5. 11,179

6. 33 miles

7. 44 miles

8. 763 meters

9. incorrect; 1948

Objective 1

1. 12

Objective 2

3. 31

Objective 3

5. 987

Objective 4

7. 1041

9. 10,415

Objective 5

11. 44 miles

Objective 6

13. correct

1.3 Subtracting Whole Numbers

Key Terms

1. minuend
2. regrouping
3. subtrahend
4. difference

Now Try

1a. $13 - 8 = 5$ or $13 - 5 = 8$ 1b. $12 - 9 = 3$ or $12 - 3 = 9$

2a. $9 = 7 + 2$ or $9 = 2 + 7$

2b. $32 = 19 + 13$ or $32 = 13 + 19$

3. minuend: 87; subtrahend: 12; difference: 75   4a. 43

4b. 111   4c. 6210   5a. correct

5b. incorrect;222   5c. incorrect; 1302   6. 39

7. 488   8. 2788   9. 5003

10a. correct   10b. incorrect; 378   11. 95 passengers

Objective 1

1. $187 - 38 = 149$; $187 - 149 = 38$ 3. $785 + 426 = 1211$

Objective 2

5. minuend: 35; subtrahend: 9; difference: 24

Objective 3

7. 6012

Objective 4

9. not correct; 153

Objective 5

11. 198   13. 6703

Objective 6

15. $263

Answers

1.4 Multiplying Whole Numbers

Key Terms

1. commutative property of multiplication

2. factors

3. associative property of multiplication

4. multiple

5. product

6. partial products

7. chain multiplication problem

Now Try

1a. 42

1b. 0

1c. 16

2. 40

3a. 258

3b. 3969

4a. 460

4b. 70,800

4c. 519,000

5a. 17,800

5b. 8400

6. 2880

7a. 28,116

7b. 8745

8a. 89,358

8b. 32,388,875

9. $5372

Objective 1

1. factors: 9, 12; 108

3. 72

Objective 2

5. 84

Objective 3

7. 216

9. 180,054

Objective 4

11. 20,000

Objective 5

13. 4416

15. 1,644,648

Objective 6

17. $9728

1.5   Dividing Whole Numbers

Key Terms

1. remainder   2. dividend   3. quotient

4. divisor   5. short division

Now Try

1a. $4\overline{)28}^{\,7}$ or $\dfrac{28}{4}=7$ 1b. $72\div 9=8$ or $9\overline{)72}^{\,8}$ 1c. $32\div 8=4$ or $\dfrac{32}{8}=4$

2a. dividend: 63; divisor: 9; quotient: 7   2b. dividend: 56; divisor: 8; quotient: 7

2c. dividend: 54; divisor: 6; quotient: 9   3a. 0

3b. 0   3c. 0   3d. 0

4a. $4\cdot 9=36$ or $9\cdot 4=36$ 4b. $3\cdot 6=18$ or $6\cdot 3=18$ 4c. $8\cdot 5=40$ or $5\cdot 8=40$

5a. undefined   5b. undefined   5c. undefined

6a. 1   6b. 1   6c. 1

7a. 12   7b. 23   7c. 97

8. 23   9. 74 R 1   10. 261 R 5

11a. correct   11b. incorrect; 78 R1   12a. Yes

12b. No   13a. Yes   13b. No

14a. Yes   14b. Yes   14c. No

15a. Yes   15b. No

Objective 1

1. $3\overline{)15}^{\,5}$; $\dfrac{15}{3}=5$

Objective 2

3. dividend: 63; divisor: 7; quotient: 9   5. dividend: 44; divisor: 11; quotient: 4

Objective 3

7. 0

Objective 4

Answers

9. undefined   11. undefined

Objective 5

13. 1

Objective 6

15. 38

Objective 7

17. 72 R 3

Objective 8

19. incorrect; 1522 R 5   21. incorrect; 6050

Objective 9

23. 2: yes; 3: no; 5: no; 10: no

1.6   Long Division

Key Terms

1. dividend   2. remainder   3. divisor

4. quotient   5. long division

Now Try

1. 73   2. 38 R 55   3. 209 R20

4a. 5   4b. 47   4c. 606

5a. 135   5b. 2006   6a. incorrect; 37 R62

6b. correct

Objective 1

1. 77   3. 654 R22

Objective 2

5. 13

Objective 3

7. correct   9. incorrect; 460 R3

1.7 Rounding Whole Numbers

Key Terms

1. front end rounding   2. rounding   3. estimate

Now Try

1a. 8<u>6</u>4 is closer to 860.   1b. <u>2</u>398 is closer to 2000.

1c. 9<u>3</u>7,645 is closer to 940,000.   2. 500

3. 25,000   4a. 20,000   4b. 837,000,000

5. 5470, 5500, 5000   6a. 190   6b. 40

6c. 1800   7a. 2700   7b. 300

7c. 480,000   8a. 17,850   8b. 8100

8c. 40,000

Objective 1

1. 257,3<u>0</u>1   3. 6<u>4</u>5,371

Objective 2

5. 17,000

Objective 3

7. $400 + 200 + 300 + 200 = 1100$; 1125

9. $900 \times 800 = 720,000$; 715,008

Objective 4

11. $300 - 50 = 250$; 264

1.8 Exponents, Roots, and Order of Operations

Key Terms

1. order of operations   2. square root   3. perfect square

Now Try

1a. exponent: 2; base 6; 36   1b. exponent: 5; base: 3; 243

2a. 3   2b. 11   2c. 12

Answers

2d. 1	3a. 20	3b. 25
3c. 1	3d. 29	4a. 32
4b. 14	4c. 33	4d. 27

Objective 1

1. exponent: 2; base: 7; 49   3. exponent: 3; base: 8; 512

Objective 2

5. 324; 324

Objective 3

7. undefined   9. 21

1.9   Reading Pictographs, Bar Graphs, and Line Graphs

Key Terms

1. line graph   2. pictograph   3. bar graph

Now Try

| 1a. Idaho | 1b. 1% | 2. 18 |
| 3a. decrease | 3b. $3,500,000 | |

Objective 1

1. 12   3. 5

Objective 2

5. 200

Objective 3

7. The net sales are increasing every year.   9. 2007

1.10 Solving Application Problems

Key Terms

1. indicator words
2. sum; increased by
3. product; times

4. quotient; per
5. difference; fewer

Now Try

1. 16 seashells
2. 13,485 calories
3. 11,017 students

4. $28,288

Objective 1

1. addition
3. subtraction

Objective 2

5. $3

Objective 3

7. $40 \times 30 = 1200$ miles; 936 miles
9. $\$700 - (\$300 + \$100 + \$30) = \$270$; $268

Answers

Chapter 2 MULTIPLYING AND DIVIDING FRACTIONS

2.1   Basics of Fractions
Key Terms
1. improper fraction   2. numerator   3. proper fraction

4. denominator

Now Try

1a. shaded: $\frac{2}{5}$; unshaded: $\frac{3}{5}$ 1b. shaded: $\frac{7}{12}$; unshaded: $\frac{5}{12}$

2. $\frac{5}{4}$ 3a. Numerator: 5; Denominator: 6

3b. Numerator: 11; Denominator: 5   4a. $\frac{4}{5}$ $\frac{6}{13}$ $\frac{17}{19}$ $\frac{1}{18}$

4b. $\frac{21}{10}$ $\frac{10}{3}$ $\frac{4}{4}$ $\frac{9}{8}$

Objective 1 In each answer, the first fraction is the shaded portion, and the second fraction is the unshaded portion.

1. $\frac{3}{8}$; $\frac{5}{8}$ 3. $\frac{7}{10}$; $\frac{3}{10}$

Objective 2
5. N: 19; D:50

Objective 3
7. improper   9. proper

2.2   Mixed Numbers
Key Terms
1. proper fraction   2. mixed number   3. improper fraction

4. whole numbers

Now Try

1. $\frac{44}{5}$ 2a. $4\frac{1}{5}$ 2b. 6

Objective 1

1. $2\frac{1}{2}, 1\frac{1}{6}$
3. $4\frac{3}{4}$

Objective 2

5. $\frac{25}{4}$

Objective 3

7. $4\frac{5}{9}$
9. $19\frac{2}{11}$

2.3   Factors

Key Terms

1. factorizations
2. composite number
3. prime factorization
4. prime number
5. factors

Now Try

1a. 1, 2, 4, 8, 16, 32
1b. 1, 2, 4, 7, 8, 14, 28, 56
2. 11, 23, 37

3a. composite
3b. not composite
3c. composite

4. $2 \cdot 2 \cdot 2 \cdot 3$
5. $2 \cdot 2 \cdot 2 \cdot 7$ or $2^3 \cdot 7$
6. $2 \cdot 3 \cdot 5^2$

7a. $2 \cdot 2 \cdot 3 \cdot 3 = 2^2 \cdot 3^2$
7b. $3 \cdot 3 \cdot 13 = 3^2 \cdot 13$
7c. $2 \cdot 2 \cdot 2 \cdot 5 = 2^3 \cdot 5$

Objective 1

1. 1, 2, 7, 14
3. 1, 2, 3, 4, 6, 8, 9, 12, 18, 24, 36, 72

Objective 2

5. composite

Objective 3

7. $2^2 \cdot 7$
9. $2 \cdot 3^2 \cdot 5^2$

Answers

2.4   Writing a Fraction in Lowest Terms
Key Terms

  1. lowest terms   2. common factor   3. equivalent fractions

Now Try

1a. yes   1b. no   2a. $\dfrac{1}{3}$

2b. $\dfrac{2}{5}$ 2c. $\dfrac{2}{3}$ 2d. $\dfrac{3}{5}$

3a. $\dfrac{3}{4}$ 3b. $\dfrac{11}{8}$ 3c. $\dfrac{1}{7}$

4a. equivalent   4b. not equivalent   4c. not equivalent

Objective 1

  1. no   3. no

Objective 2

  5. $\dfrac{2}{9}$

Objective 3

  7. $\dfrac{2\cdot2\cdot2\cdot3\cdot3}{2\cdot3\cdot3\cdot5}=\dfrac{4}{5}$ 9. $\dfrac{3\cdot5\cdot5}{2\cdot2\cdot5\cdot5\cdot5}=\dfrac{3}{20}$

Objective 4

  11. equivalent

2.5   Multiplying Fractions

Key Terms

1. common factor   2. denominator   3. multiplication shortcut

4. numerator

Now Try

1a. $\dfrac{4}{15}$ 1b. $\dfrac{5}{189}$ 1c. $\dfrac{21}{1000}$

2. $\dfrac{2}{3}$ 3a. $\dfrac{5}{9}$ 3b. $\dfrac{2}{21}$

3c. $\dfrac{20}{3}$ or $6\dfrac{2}{3}$ 3d. $\dfrac{5}{42}$ 4a. 2

4b. 10   5a. $\dfrac{1}{2}$ yd^2 5b. 10 m^2

Objective 1

1. $\dfrac{5}{12}$ 3. $\dfrac{5}{81}$

Objective 2

5. $\dfrac{2}{3}$

Objective 3

7. 42   9. 5

Objective 4

11. $2\dfrac{1}{2}$ m^2

2.6   Applications of Multiplication
Key Terms
  1. indicator words   2. product   3. reciprocals

Now Try

  1. 500 items   2. $104   3. $\frac{1}{10}$

Objective 1
  1. 320 muffins   3. 6 gallons

2.7   Dividing Fractions
Key Terms
  1. reciprocals   2. indicator words   3. quotient

Now Try

1a. $\frac{6}{1}$ 1b. $\frac{13}{4}$ 1c. $\frac{25}{12}$

1d. $\frac{1}{21}$ 2a. $1\frac{7}{9}$ 2b. $\frac{3}{4}$

3a. 80   3b. $\frac{11}{15}$ 4. 63

5. $\frac{1}{27}$

Objective 1
  1. $\frac{2}{9}$ 3. $\frac{1}{10}$

Objective 2
  5. $\frac{3}{4}$

Objective 3
  7. 24 Brownies   9. 40 trips

2.8   Multiplying and Dividing Mixed Numbers

Key Terms

1. simplify   2. round   3. mixed number

Now Try

1a. estimate: 35; exact: $36\frac{9}{10}$ 1b. estimate: 8; exact: $9\frac{17}{36}$

1c. estimate: 8; exact: $6\frac{8}{11}$ 2a. estimate: 10; exact: $8\frac{19}{28}$

2b. estimate: 5; exact: $5\frac{5}{23}$ 2c. estimate: 1; exact: $\frac{34}{35}$

3. estimate: 60 yards; exact: 65 yards

4. estimate: 17.5; exact: 16 dresses

Objective 1

1. 15; $13\frac{1}{3}$ 3. 42; $40\frac{3}{8}$

Objective 2

5. 1; $1\frac{1}{4}$

Objective 3

7. 246 yards; 248 yards   9. 8 tapes; 6 tapes

Chapter 3 ADDING AND SUBTRACTING FRACTIONS

3.1   Adding and Subtracting Like Fractions

Key Terms

1. unlike fractions   2. like fractions

Now Try

1a. like   1b. unlike   2a. $\frac{7}{9}$

2b. $\frac{4}{5}$ 3a. $\frac{2}{3}$ 3b. $1\frac{1}{5}$

Objective 1

1. unlike   3. like

Objective 2

5. $\frac{4}{5}$

Objective 3

7. $\frac{5}{14}$ 9. $2\frac{1}{2}$

3.2   Least Common Multiples

Key Terms

1. least common multiple   2. LCM

Now Try

1. 56   2. 36   3. 60

4. 105   5a. 30   5b. 30

6a. 90   6b. 72   7a. 60

7b. 60   8. $\frac{35}{40}$ 9a. $\frac{21}{45}$

9b. $\frac{27}{39}$

Objective 1

1. 14   3. 150

Objective 2

5. 70

Objective 3

7. 336   9. 400

Objective 4

11. 108

Objective 5

13. 4   15. 180

3.3   Adding and Subtracting Unlike Fractions

Key Terms

1. least common denominator   2. LCD

Now Try

1. $\dfrac{13}{16}$ 2a. $\dfrac{1}{3}$ 2b. $\dfrac{13}{18}$

3a. $\dfrac{11}{12}$ 3b. $\dfrac{23}{36}$ 4a. $\dfrac{4}{35}$

4b. $\dfrac{2}{3}$ 5a. $\dfrac{2}{15}$ 5b. $\dfrac{5}{18}$

Objective 1

1. $\dfrac{33}{40}$ 3. $\dfrac{7}{8}$

Objective 2

5. $\dfrac{29}{66}$

Objective 3

7. $\dfrac{3}{8}$ 9. $\dfrac{37}{50}$

Answers

Objective 4

11. $\dfrac{21}{80}$

3.4   Adding and Subtracting Mixed Numbers

Key Terms

1. regrouping when subtracting fractions

2. regrouping when adding fractions

Now Try

1a. $13;\ 12\dfrac{1}{2}$ 1b. $5;\ 5\dfrac{1}{8}$ 2. $21;\ 21\dfrac{1}{9}$

3a. $8;\ 8\dfrac{4}{9}$ 3b. $3;\ 3\dfrac{7}{12}$ 4a. $10\dfrac{1}{16}$

4b. $4\dfrac{13}{30}$

Objective 1

1. $8;\ 7\dfrac{71}{72}$ 3. $7;\ 6\dfrac{25}{56}$

Objective 2

5. $31;\ 30\dfrac{11}{15}$

Objective 3

7. $8\dfrac{1}{6}$ 9. $1\dfrac{49}{72}$

3.5   Order Relations and the Order of Operations

Key Terms

1. $<$ 2. $>$

Now Try

1a. $\dfrac{1}{8} < \dfrac{8}{3}$ 1b. $\dfrac{13}{4} > 1$ 1c. $\dfrac{9}{2} < \dfrac{21}{4}$

2a. $\dfrac{5}{6}$ 2b. $\dfrac{21}{8}$ 3a. $\dfrac{1}{64}$

3b. $\dfrac{36}{49}$ 3c. $\dfrac{9}{25}$ 4a. $\dfrac{10}{27}$

4b. $\dfrac{3}{50}$ 4c. $\dfrac{27}{40}$

Objective 1

1. $>$ 3. $<$

Objective 2

5. $5\dfrac{1}{16}$

Objective 3

7. $\dfrac{4}{15}$ 9. $1\dfrac{1}{6}$

Answers

Chapter 4 DECIMALS

4.1   Reading and Writing Decimals

Key Terms

  1. decimals   2. place value   3. decimal point

Now Try

  1a. 0.6   1b. 0.05   1c. 0.37

  1d. 0.007   1e. 0.049   1f. 0.518

  2a. 8 hundreds; 6 tens; 2 ones; 9 tenths; 3 hundredths

  2b. 0 ones; 0 tenths; 0 hundredths; 7 thousandths; 6 ten-thousandths;
      9 hundred-thousandths

  3a. nine tenths   3b. fifty-three hundredths   3c. seven hundredths

  3d. five hundred two thousandths

  3e. four hundred sixty-nine ten-thousandths   4a. four and seven tenths

  4b. eighteen and nine thousandths   4c. eight-two ten-thousandths

  4d. fifty-seven and nine hundred six thousandths   5a. $\dfrac{27}{100}$

  5b. $\dfrac{303}{1000}$ 5c. $3\dfrac{2636}{10,000}$ 6a. $\dfrac{1}{5}$

  6b. $\dfrac{7}{20}$ 6c. $6\dfrac{1}{25}$ 6d. $562\dfrac{101}{2500}$

Objective 1

  1. $\dfrac{8}{10}$; 0.8; eight tenths   3. $\dfrac{58}{100}$; 0.58; fifty-eight hundredths

Objective 2

  5. 7; 1   6. tens; ones; tenths; hundredths; thousandths

Objective 3

  7. eight hundredths   9. 38.00052

Objective 4

  11. $3\dfrac{3}{5}$

390   Copyright © 2014 Pearson Education, Inc.

4.2   Rounding Decimal Numbers

Key Terms

1. decimal places   2. rounding

Now Try

1. 43.803	2a. 0.80	2b. 7.78
2c. 22.040	2d. 0.6	3a. $2.08
3b. $425.10	4a. $38	4b. $307
4c. $881	4d. $6860	4e. $1

Objective 1

1. up

Objective 2

3. 489.8   5. 989.990

Objective 3

7. $11,840

4.3   Adding and Subtracting Decimal Numbers

Key Terms

1. front end rounding   2. estimating

Now Try

1a. 15.630	1b. 21.709	2a. 36.844
2b. 1013.931	3a. 11.612	3b. 10.57
4a. 4.654	4b. 3.26	4c. 0.501
5a. 14; 13.847	5b. $44; $42.95	5c. 20 ft; 19.142 ft
5d. 40; 35.154		

Objective 1

1. 92.49   3. 72.453

Objective 2

5. 42.566

Answers

Objective 3

7. 680; 676.60   9. 5; 4.838

4.4   Multiplying Decimal Numbers

Key Terms

1. factor   2. decimal places   3. product

Now Try

1. 10.793   2. 0.00186   3. 100; 112.8229

Objective 1

1. 90.71   3. 0.0037

Objective 2

5. 600; 756.6478

4.5   Dividing Decimal Numbers

Key Terms

1. dividend   2. repeating decimal   3. quotient

4. divisor

Now Try

1a. 1.413   1b. 15.03   2. 20.178

3. 16.791   4a. 20,410   4b. 1.78

5. 40; 37.8   6a. 12.18   6b. 22.62

6c. 0.373

Objective 1

1. 6.96   3. 2.359

Objective 2

5. 128.25

Objective 3

7. reasonable   9. reasonable

Objective 4

11. 53.24

4.6   Fractions as Decimals

Key Terms

1. equivalent   2. numerator   3. denominator

4. mixed number

Now Try

1a. 0.0625   1b. 4.375   2. 0.417

3a. >   3b. =   3c. <

3d. =   4a. 0.3, 0.3057, 0.307   4b. $\frac{1}{5}$, $\frac{2}{9}$, 0.23, $\frac{3}{13}$

Objective 1

1. 0.125   3. 19.708

Objective 2

5. $\frac{3}{11}$, 0.29, $\frac{1}{3}$

Chapter 5 RATIO AND PROPORTION

5.1   Ratios

Key Terms

1. ratio   2. numerator; denominator

Now Try

1a. $\dfrac{17}{8}$ 1b. $\dfrac{8}{5}$ 2a. $\dfrac{2}{3}$

2b. $\dfrac{2}{1}$ 2c. $\dfrac{2}{3}$ 3. $\dfrac{3}{8}$

4a. $\dfrac{19}{9}$ 4b. $\dfrac{22}{17}$ 5. $\dfrac{5}{7}$

Objective 1

1. $\dfrac{25}{19}$ 3. $\dfrac{1}{4}$

Objective 2

5. $\dfrac{9}{2}$

Objective 3

7. $\dfrac{2}{7}$ 9. $\dfrac{5}{6}$

5.2   Rates

Key Terms

1. unit rate   2. cost per unit   3. rate

Now Try

1a. $\dfrac{1 \text{ dollar}}{5 \text{ pages}}$ 1b. $\dfrac{15 \text{ strokes}}{1 \text{ minute}}$ 1c. $\dfrac{33 \text{ strawberries}}{2 \text{ cakes}}$

2a. 28 miles/gallon   2b. $1.26/pound   2c. $145/day

3. 5 pints at $1.70/pint   4a. eight-pack   4b. Brand T

Objective 1

1. $\dfrac{7\text{ pills}}{1\text{ patient}}$ 3. $\dfrac{32\text{ pages}}{1\text{ chapter}}$

Objective 2

5. \$225/pound

Objective 3

7. 24 ounces for \$2.08   9. 5 cans for \$2.75

5.3   Proportions

Key Terms

1. proportion   2. cross products

Now Try

1a. $\dfrac{24}{17}=\dfrac{72}{51}$ 1b. $\dfrac{\$10}{7\text{ cans}}=\dfrac{\$60}{42\text{ cans}}$ 2a. $\dfrac{9}{7}\neq\dfrac{4}{3}$, false

2b. $\dfrac{1}{3}=\dfrac{1}{3}$, true   3a. $306=306$, true   3b. $32\neq35$, false

Objective 1

1. $\dfrac{50}{8}=\dfrac{75}{12}$ 3. $\dfrac{3}{33}=\dfrac{12}{132}$

Objective 2

5. $\dfrac{6}{5}=\dfrac{6}{5}$; true

Objective 3

7. $2100=2450$; false   9. $12.814=12.07$; false

Answers

5.4   Solving Proportions
Key Terms

1. proportion   2. cross products   3. ratio

Now Try

1a. 32   1b. 10.67   2a. $2\frac{3}{5}$

2b. 6.4

Objective 1

1. 36   3. 40

Objective 2

5. 21

5.5   Solving Application Problems with Proportions
Key Terms

1. ratio   2. rate

Now Try

1. $\dfrac{23\ \text{hr}}{4\ \text{apt}} = \dfrac{x\ \text{hr}}{16\ \text{apt}}$; $x = 92$ hr   2. $\dfrac{4}{5} = \dfrac{x}{540}$; $x = 432$ people

Objective 1

1. $108   3. approximately 333 deer

Chapter 6 PERCENTS

6.1   Basics of Percent
Key Terms

  1. ratio   2. percent   3. decimals

Now Try

1a. 9%	1b. 73%	2a. 0.29
2b. 0.86	3a. 0.48	3b. 2.40 or 2.4
3c. 0.025	3d. 0.008	4a. 43%
4b. 230%	4c. 75.1%	5a. $35
5b. 172 miles	5c. 156 glasses	6a. 20 students
6b. 27 photographs	6c. 5 miles	

Objective 1

  1. 68%   3. 45%

Objective 2

  5. 3.10

Objective 3

  7. 20%   9. 493%

Objective 4

11. $1040

Objective 5

13. 125 signs   15. $0.98 or 98¢

6.2   Percents and Fractions
Key Terms

  1. lowest terms   2. percent

Now Try

1a. $\frac{3}{10}$	1b. $\frac{39}{50}$	1c. $1\frac{3}{4}$

Answers

2a. $\dfrac{109}{250}$ 2b. $\dfrac{2}{9}$ 3a. 35%

3b. 67.5%   3c. 38.9%   4a. 12.5%

4b. $\dfrac{2}{5}$ 4c. 25%

Objective 1

1. $1\dfrac{2}{5}$ 3. $\dfrac{139}{250}$

Objective 2

5. 27.5%

Objective 3

7. 0.375; 37.5%   9. $\dfrac{13}{40}$; 32.5%

6.3   Using the Percent Proportion and Identifying the Components in a Percent Problem

Key Terms

1. whole   2. part   3. percent proportion

Now Try

1a. whole   1b. part   1c. percent

2a. 800   2b. 5%   2c. 3.5

3a. 59   3b. 7.75   3c. unknown

4a. 650   4b. unknown   4c. 2650

5a. $\dfrac{\text{unknown}}{650} = \dfrac{7.75}{100}$ 5b. $\dfrac{7470}{9000} = \dfrac{83}{100}$ 5c. $\dfrac{343}{\text{unknown}} = \dfrac{35}{100}$

Objective 1

1. $\dfrac{\text{part}}{\text{whole}} = \dfrac{\text{percent}}{100}$

Objective 2

3. 87.5

Objective 3

5. 83%   7. 17%

Objective 4

9. 487

Objective 5

11. 29.81   13. unknown

6.4   Using Proportions to Solve Percent Problems

Key Terms

1. cross products   2. percent proportion

Now Try

1. 2470   2a. 3040 pens   2b. 93 miles

2c. 122.5 gallons   2d. $2.56   3. 99 cars

4a. 1450 cups of coffee   4b. 950 miles   5. 150 students

6a. 12.5%   6b. 5000%   7. 2.7%

8. 121%

Objective 1

1. 280   3. 946 drivers

Objective 2

5. 500

Objective 3

7. 2.5%   9. 85%

6.5   Using the Percent Equation

Key Terms

1. percent   2. percent equation

Now Try

1a. $63   1b. 91 packages   1c. 186 students

Answers

2a. 120 units   2b. 2491 points   2c. 60

3a. 29%   3b. 20%   3c. 320%

3d. 0.4%

Objective 1

1. 21.6   3. 1.4

Objective 2

5. 1500

Objective 3

7. 20%   9. 244.4%

6.6   Solving Application Problems with Percent

Key Terms

1. commission   2. percent of increase or decrease

3. sales tax   4. discount

Now Try

1. $810   2. 1.5%   3. $4680

4. 15%   5. $9.98   6. 20%

7. 2.9%

Objective 1

1. $3.50; $53.50   3. 5.5%

Objective 2

5. 15%

Objective 3

7. $9.75

Objective 4

9. 28.3%   11. 11.4%

6.7   Simple Interest

Key Terms

1. rate of interest
2. interest
3. interest formula
4. simple interest
5. principal

Now Try

1. $4
2. $124
3. $29.75
4. $1240

Objective 1

1. $475.20
3. $588

Objective 2

5. $3075

6.8   Compound Interest

Key Terms

1. compound interest
2. compound amount
3. compounding

Now Try

1. $4867.20
2. $3595.52
3a. $1.23
3b. $1.62
4a. $2024.48, $424.48
4b. $26,624.80, $3624.80

Objective 1; Objective 2

1. $2100
3. $2315.25; $315.25

Objective 3

5. $3663.68

Objective 4

7. $1.50
9. $1.90

Objective 5

11. $34,730.06; $13,330.06

Chapter 7 MEASUREMENT

7.1   Problem Solving with U.S. Customary Measurements

Key Terms

1. metric system   2. unit fractions

3. U.S. measurement units

Now Try

1a. 1 lb

1b. 4 qt

2a. 210 inches

2b. 56 ounces

2c. 1.25 or $1\frac{1}{4}$ minutes

2d. $6.3\overline{3}$ or $6\frac{1}{3}$ hours

3a. 432 inches

3b. $\frac{1}{3}$ ft

4a. 21,120 feet

4b. $1\frac{1}{2}$ tons

4c. 13 quarts

5a. $\frac{3}{4}$ gal

5b. 7200 minutes

6a. $6.36 per pound

6b. 3.5 or $3\frac{1}{2}$ tons

Objective 1

1. 2000

3. 8

Objective 2

5. 5 gallons

Objective 3

7. 3.5 or $3\frac{1}{2}$ gallons

9. 3.75 or $3\frac{3}{4}$ pounds

Objective 4

11. $26\frac{1}{4}$ qt

7.2   The Metric System—Length

Key Terms

1. prefix   2. metric conversion line   3. meter

Now Try

1a. mm	1b. km	1c. cm
2a. 5.4 km	2b. 76 mm	3a. 675 mm
3b. 0.9865 km	3c. 431 cm	4a. 23,000 mm
4b. 0.042 m	4c. 0.95 km	

Objective 1

1. m   3. km

Objective 2

5. 0.45 km

Objective 3

7. 19.4 mm   9. 0.000035 cm

7.3   The Metric System—Capacity and Weight (Mass)

Key Terms

1. gram   2. liter

Now Try

1a. 5 mL	1b. 100 L	2a. 0.973 L
2b. 3850 mL	3a. 48 kg	3b. 590 g
3c. 3 mg	4a. 3720 mg	4b. 0.084 kg
5a. 125 mL	5b. 1.19 kg	5c. 400 m

Objective 1

1. mL   3. mL

Objective 2

5. 836,000 L

403

Answers

Objective 3

  7. mg   9. g

Objective 4

 11. 760 g

Objective 5

 13. L   15. cm

7.4   Problem Solving with Metric Measurement

Key Terms

  1. gram   2. meter   3. liter

Now Try

  1. 1.54 kg   2. 200 g   3. 5600 mL

Objective 1

  1. $12.99   3. 12 pills

7.5   Metric–U.S. Measurement Conversions and Temperature

Key Terms

  1. Celsius   2. Fahrenheit

Now Try

  1. 42.3 ft   2a. 21.3 lb   2b. 75.8 L

 3a. 65°C   3b. 4°C   4. 50°C

  5. 302°F

Objective 1

  1. 468.5 km   3. 737.1 g

Objective 2

  5. 37°C

Objective 3

  7. 17°C   9. 204°C

Chapter 8 GEOMETRY

8.1   Basic Geometric Terms

Key Terms

1. ray
2. perpendicular lines
3. obtuse angle

4. point
5. angle
6. line

7. acute angle
8. degrees
9. straight angle

10. parallel lines
11. line segment
12. intersecting lines

13. right angle

Now Try

1a. line segment, $\overline{EF}$
1b. ray, $\overrightarrow{CD}$
1c. line, $\overleftrightarrow{RS}$

2a. intersecting
2b. parallel
3. $\angle VSW$ or $\angle WSV$

4a. straight
4b. obtuse
4c. acute

4d. right
5a. intersecting
5b. perpendicular

Objective 1

1. ray, $\overrightarrow{CB}$
3. line segment, $\overline{KL}$

Objective 2

5. parallel

Objective 3

7. $\angle COD$
9. $\angle MON$

Objective 4

11. obtuse

Objective 5

13. perpendicular
15. intersecting

Answers

8.2   Angles and Their Relationships

Key Terms

  1. vertical angles   2. congruent angles   3. supplementary angles

  4. complementary angles

Now Try

  1. $\angle SRT$ and $\angle TRU$; $\angle URV$ and $\angle VRW$ 2a. $18°$

2b. $78°$

  3. $\angle OQP$ and $\angle PQN$; $\angle RST$ and $\angle BMC$; $\angle OQP$ and $\angle BMC$, $\angle PQN$ and $\angle RST$

4a. $98°$ 4b. $12°$

  5. $\angle RPS \cong \angle QPT$ and $\angle QPR \cong \angle TPS$

  6. $\angle MKN$ and $\angle QKP$; $\angle MKQ$ and $\angle NKP$ 7a. $33°$

7b. $105°$ 7c. $42°$ 7d. $42°$

Objective 1

  1. $47°$

  3. $\angle LKM$ and $\angle MKN$; $\angle MKN$ and $\angle NKO$; $\angle NKO$ and $\angle OKL$; $\angle OKL$ and $\angle LKM$

Objective 2

  5. $\angle VLB$ and $\angle CLW$; $\angle VLC$ and $\angle BLW$

8.3   Rectangles and Squares

Key Terms

  1. area   2. rectangle   3. perimeter

  4. square

Now Try

  1. 22 ft   2. 30 ft^2 3a. 32 m

3b. 64 m^2 4a. $P = 24$ cm   4b. $A = 31 \text{ cm}^2$

Objective 1

  1. $P = 58$ in.; $A = 204 \text{ in.}^2$ 3. $P = 281.2$ cm; $A = 3859.68 \text{ cm}^2$

Objective 2

5. $P = 32.8$ km; $A = 67.24$ km^2

Objective 3

7. $P = 30$ ft; $A = 18$ ft^2 9. $P = 188$ ft; $A = 1348$ ft^2

8.4   Parallelograms and Trapezoids

Key Terms

1. parallelogram   2. trapezoid   3. perimeter

4. area

Now Try

1. 26 in.   2a. 713 yd^2 2b. 19.22 yd^2

3. $53\frac{3}{4}$ yd   4. 1943.7 cm^2 5. 3515.4 cm^2

6. $7579

Objective 1

1. 168 m   3. 310 ft^2

Objective 2

5. 1106 m^2

8.5   Triangles

Key Terms

1. triangle   2. base   3. height

Now Try

1. 59 cm   2a. $\frac{77}{128}$ in.^{2} 2b. 28 yd^2

2c. 232.56 cm^2 3. 534 m^2 4. $4.80; $42.72

5a. 80°   5b. 65°

Objective 1

1. 37.2 ft   3. 3 in.

Answers

Objective 2

5. 112 cm^2

Objective 3

7. 81°

9. 28°

8.6   Circles

Key Terms

1. radius

2. circumference

3. circle

4. π (pi)

5. diameter

Now Try

1a. 86 m

1b. 29.5 in.

2a. 138.2 yd

2b. 23.2 m

3a. 43.0 m^2

3b. 1519.8 yd^2

4. 127.2 ft^2

5. $12.56

6. $1.96

7a. millimeter; millipede

Objective 1

1. 4 ft

3. $6\frac{1}{4}$ yd

Objective 2

5. 28.3 yd

Objective 3

7. 22.3 yd^2

9. 1061.3 m^2

Objective 4 *Other answers are possible.*

11. polynomial; polyglot

8.7   Volume

Key Terms

1. rectangular solid
2. cylinder
3. sphere
4. volume
5. pyramid
6. cone

Now Try

1. 48 m^3
2. 4.2 m^3
3. 0.9 in.^3
4. 1846.3 in.^3
5. 7536 cm^3
6. 25.8 ft^3

Objective 1

1. 2744 in.^3
3. 176 in.^3

Objective 2

5. 3267.5 ft^3

Objective 3

7. 0.1 km^3
9. 490.6 m^3

Objective 4

11. 333.3 m^3

8.8   Pythagorean Theorem

Key Terms

1. right triangle
2. hypotenuse
3. legs

Now Try

1a. 4.47
1b. 9.59
1c. 11.58
2a. 9.8 in.
2b. 12 km
3. 9.5 ft

Objective 1

1. 4.123
3. 10.100

Objective 2

5. 2.2 cm

Objective 3

7. 10.3 ft
9. 8 ft

Answers

8.9   Similar Triangles

Key Terms

1. congruent   2. similar triangles

Now Try

1. $\overline{AB}$ and $\overline{PQ}$; $\overline{AC}$ and $\overline{PR}$; $\overline{BC}$ and $\overline{QR}$;
   $\angle A$ and $\angle P$; $\angle B$ and $\angle Q$; $\angle C$ and $\angle R$

2. 12.75 yd   3. 24   4. 21 ft

Objective 1

1. $\overline{PN}$ and $\overline{SR}$; $\overline{NM}$ and $\overline{RQ}$; $\overline{MP}$ and $\overline{QS}$;
   $\angle P$ and $\angle S$; $\angle N$ and $\angle R$; $\angle M$ and $\angle Q$

3. $\overline{HK}$ and $\overline{RS}$; $\overline{GH}$ and $\overline{TR}$; $\overline{GK}$ and $\overline{TS}$;
   $\angle H$ and $\angle R$; $\angle G$ and $\angle T$; $\angle K$ and $\angle S$

Objective 2

5. $m = 90$; $r = 21$

Objective 3

7. 36 m   9. 10 m

Chapter 9 BASIC ALGEBRA

9.1   Signed Numbers

Key Terms

1. signed numbers   2. opposite of a number

3. absolute value   4. negative numbers

Now Try

1.

2a. $2 < 5$ 2b. $-8 < -3$

2c. $5 > -5$ 2d. $-9 < 0$ 3a. 6   3b. 6

3c. 0   3d. -8 4a. -6 4b. -11

4c. $-\dfrac{15}{17}$ 4d. 0   5a. 14   5b. 27

5c. $\dfrac{7}{8}$

Objective 1

1. $+17$ 3. -7

Objective 2

5.

Objective 3

7. $<$ 9. $<$

Objective 4

11. -8.23

Objective 5

13. -3 15. $\dfrac{2}{3}$

Answers

9.2 Adding and Subtracting Signed Numbers

Key Terms

1. absolute value   2. additive inverse

Now Try

1a. 3   1b. –3   1c. –9   2a. –19

2b. –42   2c. –108   2d. $-\dfrac{3}{2}$ 3a. 3

3b. –12   3c. –15   3d. 14   3e. $\dfrac{1}{30}$

4a. –8; 0   4b. 25; 0   4c. –6.9; 0   4d. 300; 0

4e. $\dfrac{8}{5}$; 0   4f. 0; 0   5a. –5   5b. –32

5c. 5   5d. 21   5e. –2   6a. 25.1

6b. –18.5   6c. $-\dfrac{11}{20}$ 6d. $-\dfrac{1}{24}$ 7a. –25

7b. 11   7c. 17   7d. –9.3

Objective 1

1. 15   3. –8

Objective 2

5. –7.9

Objective 3

7. 15   9. –281

Objective 4

11. $-\dfrac{1}{8}$

Objective 5

13. –14   15. 9.5

9.3 Multiplying and Dividing Signed Numbers

Key Terms

1. factors 2. quotient 3. product

Now Try

1a. –70 1b. –63 1c. –72

1d. –176 2a. 52 2b. 90

2c. 72 2d. 15 2e. 84

3a. –3 3b. 4 3c. 25

3d. undefined 3e. 0 3f. –3

3g. $\dfrac{3}{2}$

Objective 1

1. –39 3. –11

Objective 2

5. 4.55

9.4 Order of Operations

Key Terms

1. order of operations 2. base 3. exponent

Now Try

1. 4 2a. –36 2b. –31

3a. 13 3b. 129 3c. $-\dfrac{1}{20}$

4. –4

Objective 1

1. –9 3. –3

Objective 2

5. $-\dfrac{38}{63}$

Answers

Objective 3

7. $\dfrac{37}{10}$ or $3\dfrac{7}{10}$ 9. -1

9.5   Evaluating Expressions and Formulas

Key Terms

1. expression   2. variable

Now Try

1. -3 2. 1   3. 15

4. 63 ft^2

Objective 1

1. expression   3. expression

Objective 2

5. -21

9.6   Solving Equations

Key Terms

1. equation   2. multiplication property of equality

3. solution   4. addition property of equality

Now Try

1a. solution   1b. not a solution   2a. $n = 16$

2b. $x = -4$ 3a. $p = 5$ 3b. $r = -4$

3c. $m = 4$ 4a. $x = 35$ 4b. $r = -30$

Objective 1

1. not a solution   3. solution

Objective 2

5. $a = -2$

Objective 3

7. $m = -1.1$ 9. $x = -32$

9.7   Solving Equations with Several Steps

Key Terms

1. like terms   2. distributive property

Now Try

1. $p = -4$ 2a. 117   2b. $-2k - 10$

2c. $7y - 56$ 2d. $-8x + 32$ 3a. $23k$

3b. $-2m$ 3c. $-11x$ 4a. $y = 5$

4b. $p = 4$ 5. $m = 1$

Objective 1

1. $y = 11$ 3. $p = -2.7$

Objective 2

5. $-32 + 4x$

Objective 3

7. $5z$ 9. $2.3a$

Objective 4

11. $z = 3$

9.8   Using Equations to Solve Application Problems
Key Terms

1. indicator words   2. sum; increased by   3. product; double

4. quotient; per   5. difference; less than

Now Try

1a. $x - 16$ 1b. $x + 19$ 1c. $20 - x$

1d. $12 + x$ 1e. $x + 13$ 1f. $x - 27$

1g. $-54 + x$ 1h. $x + 99$ 1i. $38 - x$

2a. $-7x$ 2b. $5x$ 2c. $5x$

2d. $\dfrac{16}{x}$ 2e. $\dfrac{x}{24}$ 2f. $\dfrac{-11}{x}$

2g. $\dfrac{1}{5}x$ or $\dfrac{x}{5}$ 2h. $\dfrac{x}{6}$ 3. $x = 9$

4. 12   5. -3 6. 29 books

7. 8 cm

Objective 1

1. $9 + x$ 3. $3x$

Objective 2

5. $45 - 2x = 35$; $x = 5$

Objective 3

7. \$14   9. -5

Chapter 10 STATISTICS

10.1   Circle Graphs

Key Terms

1. circle graph

2. protractor

Now Try

1. $\dfrac{3}{58}$

2. $\dfrac{5}{9}$

3. $142,500

4a. mysteries: 108°; biographies: 54°; cookbooks: 36°; romance novels: 90°; science: 54°; business: 18°

4b.

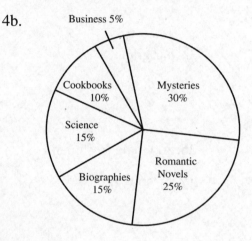

Objective 1

1. $10,400

3. $400

Objective 2

5. $\dfrac{1000}{1800}$ or $\dfrac{5}{9}$

Objective 3

7. parts: 5%; hand tools: 20%; bench tools: 25%; brass fittings: 35%; hardware: 15%

Answers

9.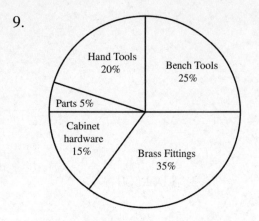

10.2   Bar Graphs and Line Graphs

Key Terms

1. line graph   2. double-bar graph   3. bar graph

4. comparison line graph

Now Try

1. 1800 students   2a. 250 female seniors   2b. 500 male freshmen

3a. $60   3b. December   4a. $1,500,000

4b. $4,000,000

Objective 1

1. 1600 students   3. 400 students

Objective 2

5. sophomores

Objective 3

7. September   9. $30

Objective 4

11. $2,000,000

10.3   Frequency Distributions and Histograms

Key Terms

1. histogram   2. frequency distribution

Now Try

1.

Scores	Tally	Frequency
92					3
98				2
102						5
106					3
112						4
121			1

2a. 17 students   2b. 18 students   3. 5 students

Objective 1

1. |||; 3   3. |; 1

Objective 2

5. 5 patients

Objective 3

7. 13 patients   9. 4 patients

Answers

10.4   Mean, Median, and Mode
Key Terms
1. dispersion   2. mode   3. weighted mean

4. mean   5. median   6. bimodal

7. range

Now Try
1. 272   2. 52.7   3. 40.2

4. 2.5   5. 18   6. 0.06

7a. 3   7b. 13 and 16   7c. no mode

Objective 1
1. 60.3   3. 273.1

Objective 2
5. 4.7

Objective 3
7. 232   9. 239.5

Objective 4
11. 24, 35, 39